Hydrogen Sulfide (H_2S) Hazards

Hydrogen Sulfide (H_2S) Hazards

Risk Recognition and Prevention

M. Darryl Yoes

Safety Consulting International, LLC
Jackson, LA, USA

WILEY

Published by John Wiley & Sons, Inc., Hoboken, New Jersey.

For general information on our other products and services or for technical support, please contact our Customer Care Department within the United States at (800) 762-2974, outside the United States at (317) 572-3993 or fax (317) 572-4002.

Wiley also publishes its books in a variety of electronic formats. Some content that appears in print may not be available in electronic formats. For more information about Wiley products, visit our web site at www.wiley.com.

Library of Congress Cataloging-in-Publication Data has been applied for:

Print ISBN: 9781394376520
ePDF ISBN: 9781394376544
ePub ISBN: 9781394376537
oBook ISBN: 9781394376551

Cover Design: Wiley
Cover Image: © Karl Hendon/Getty Images

U.S Chemical Safety and Hazard Investigation Board (CSB)/ Public domain

U.S Chemical Safety and Hazard Investigation Board (CSB)/Public domain

1. Wear your H_2S Detector anytime you are working near H_2S.
2. Immediately leave the area if the detector alarms.
3. Always gas test before entering a confined space.
4. Never enter an excavation without a gas test.
5. Verify the correct flange when opening process equipment.
6. Wear an SCBA when working near H_2S.
7. Use fall protection when opening a tank truck or railcar hatch.
8. Never attempt rescue unless you are trained and have proper rescue equipment.
9. Never mix a sulfide with an acid.
10. Always be aware of and respect H_2S.

This book focuses on the potential exposure to hydrogen sulfide, its hazards, effects, and methods of prevention.

According to the Bureau of Labor Statistics, hydrogen sulfide is one of the leading causes of workplace gas inhalation deaths in the United States and ranks first in the oil and gas industry. Included are case studies, what happened, where it happened, and why, and most importantly, how it can be prevented. Many other H_2S incidents are reported in the OHSA database.

Contents

Preface

Unfortunately, incidents involving personnel exposure to hydrogen sulfide and fatalities, or even multiple fatalities, are still happening today, even though you may not hear of them. As an example, a tragic hydrogen sulfide incident happened at a refinery East of Houston on October 10, 2024, involving two fatalities and multiple injuries. I recently reread the Final Investigation Report by the US Chemical Safety and Hazard Investigation Board (CSB) on the Hydrogen Sulfide Release at the Aghorn Waterflood Station in Odessa, Texas, which occurred on October 26, 2019. If you are not familiar, the operator at Aghorn was killed by hydrogen sulfide while checking a pump alarm. Unfortunately, his wife went to check on her husband after not hearing from him, and she was also very tragically killed by exposure to H_2S. The operator and his wife both died due to hydrogen sulfide while their young children were waiting for them in the wife's vehicle parked just outside the husband's pumphouse. I knew about this incident, having read the report previously. Still, this incident was tragic. I felt that the lessons learned from this and other similar incidents in our industry and other industries should be more broadly shared, especially the valuable lessons learned. As you will learn in this book, there have been many other similar incidents that have happened, not only in the United States but worldwide, many of them involving the loss of life and sometimes several lives from the same release. Unfortunately, as in the case at Aghorn, other people are killed as they rush in to help save the first victim and are also killed by this deadly gas.

Due to the October 10, 2024 hydrogen sulfide incident at the Pemex facility, the facility was evacuated, and the cities of Deer Park, North Pasadena, and surrounding communities were required to shelter in place. We know from the media reports that the incident at the Pemex Refinery occurred when workers opened flanges on a long and continuous pipeline. This initial report was updated a few weeks later with information that the workers were in the process of installing isolation blinds in the pipeline and somehow got off onto the wrong pipeline. They reportedly opened a flange on a pipeline carrying pressurized high-concentration hydrogen sulfide on an amine regenerator unit. Amine units process hydrogen sulfide in concentrations of 80–90%, which can kill in less than a minute. Pemex reported that there were 2 fatalities and 13 hospitalized because of the incident. Local agencies reported that 35 plant victims were treated at the site for exposure to hydrogen sulfide. The US Chemical Safety and Hazard Investigation Board (CSB) dispatched a team of investigators to determine what happened, and we will have a detailed report when that investigation is complete. From my perspective, there must be more focus anytime workers are working on or near piping systems carrying high-concentration hydrogen sulfide and other similar toxics or asphyxiants, and that is precisely what this book is all about. I will talk more about the Pemex incident in the case study section of this book and how incidents like this can be avoided.

The Bureau of Labor Statistics reported that between 2011 and 2017, 46 workers in the United States died due to exposure to hydrogen sulfide gas. I believe that the number may even be higher. When you recognize how H_2S deaths occur, some may not be reported to OSHA. They happen in oil and gas operations, sewage treatment, ore mining, and pulp and paper operations, which are likely reported. However, deaths from H_2S also occur on dairy farms, cattle and pork ranch operations, agricultural farms, grass bogs, hazardous waste sites, etc. I'm concerned that some of these may go unreported.

In the appendices, you will find a summary of over 30 years of hydrogen sulfide incidents in the United States. This includes many deaths due to exposure to H_2S vapors. Since H_2S is also flammable, several significant explosions and fires have occurred due to the presence of H_2S in the workplace, homes, and businesses. I think for most readers, the sheer number of fatalities related to hydrogen sulfide will be an eye-opener. In many petroleum refineries and petrochemical plants, I believe H_2S has become taken for granted as a routine part of the industry. We have forgotten how deadly it can be, especially in confined spaces or similar areas with little or no ventilation. Unfortunately, as illustrated earlier, the incident that occurred at Pemex tells us that these incidents are still happening today, and H_2S is just as deadly as it ever has been. It takes only one mistake or omission, and people's lives at the facility and nearby communities are at risk.

There should be no doubt that hydrogen sulfide (H_2S) is a deadly chemical when in higher concentrations. One factor that makes it so deadly is that it also deadens the olfactory nerves in higher concentrations so that the victim can no longer smell the unique rotten egg odor. They are led to believe it has dissipated and is no longer present. I talk about this in the first Process Safety Book, *Process Operations Safety: The What, Why, and How Behind Safe Petrochemical Plant Operations*, but we will do a deeper dive in this book about the hazards and the effects. We will also discuss multiple case studies and, most importantly, how the hazards can be identified and how they can be mitigated. The focus of this book is to highlight the hazards and to prevent future incidents through education and by improving the effectiveness of our procedures and work practices.

Hydrogen sulfide can usually be detected in low concentrations by a unique odor, which smells like rotten eggs. It can be detected at concentrations as low as 0.01–1.5 ppm. However, when the concentration reaches about 100–150 ppm, the olfactory nerves, the nerves responsible for the sense of smell, are deadened, and the victim can no longer detect its presence by the sense of smell. At this concentration, the victim or victims automatically assume the hazard has gone away, and unfortunately, they quickly collapse and can die from exposure to hydrogen sulfide.

The NIOSH concentration for "Immediately Dangerous to Life and Health" (IDLH) is only 100 ppm. And, at about 1,000–2,000 ppm, death is essentially instantaneous. In most petroleum refineries, we have streams circulating with H_2S concentrations many times these levels, including some streams reaching 80–90% and higher (equivalent to 800,000–900,000 ppm). This is truly incredible when you realize that the limit for IDLH is only 100 ppm or 0.0125% of that which may circulate in refinery systems.

In this book, we will discuss the hazards in more detail and review several important case studies and the key lessons learned in each one to learn more about how an incident like these can be prevented at your facility. I will be the first to tell you that you will notice that some of the causal factors and precautions are repeated more than once as you go through the material. This is not by accident. I feel these are so important, and it takes only one omission and a life, or even several lives, can be lost. Therefore, the causal factors and how they can be prevented were well worth repeating.

I also encourage you to review Appendix 2, "Documented Cases of Hydrogen Sulfide Poisoning," which provides a good overview of the hydrogen sulfide incident, fatality, and injury data from the OSHA database. The Appendix contains 30 years of H_2S incident data and there is much to learn here.

The information provided in the book will help improve safety performance in the petroleum and petrochemical industry and all other industries that use or generate hydrogen sulfide in gas or liquid form. However, neither Safety Consulting International, LLC, the author M. Darryl Yoes, nor the company producing the documents contained herein warrants or represents, expressly or by implication, the correctness or accuracy of the content or the information presented here. This material is presented to improve operations' safety worldwide and is not intended to be used as operating guidance or procedures. Any use of the material contained herein is done with the user accepting legal liability or responsibility for the consequence of its use or misuse.

Safety Consulting International, LLC, the author, or any other company or person as outlined earlier, make any claim, representation, or warranty, expressed or implied, that acting by this book or its contents will produce any particular results with regard to the subject matter contained herein, or satisfy the requirements of any applicable federal, state, or local laws and regulations. Nothing in this document constitutes technical advice. If such advice is required, it should be sought from an attorney, a qualified company, or a qualified individual.

No material known to be copyrighted or proprietary has been intentionally used in this document without the owner's express approval or permission. The author has made extraordinary efforts to contact all owners of images used in this book to secure consent to use photographs. All information presented is readily available in the public sector or represents the author's experience and over 50 years of collecting process safety information. The copyright owner approved most images; the author took those without attribution. There may be several where, after an extensive search, the original could not be found; these are noted as such.

Notes and Credits

The cover photograph is credited to the US Chemical Safety and Hazard Investigation Board (CSB) and is from the CSB Final Report on the Aghorn Hydrogen Sulfide Fatality Report. An operator and his wife lost their lives here due to exposure to hydrogen sulfide.

The hydrogen sulfide warning sign used on the cover is from the US Occupational Safety and Health Administration (OSHA) (Hydrogen Sulfide: Evaluating and Controlling Exposure) (osha.gov).

Additional Resources

US Occupational Safety and Health Administration (OSHA), Hydrogen Sulfide – Hazards and Health Hazards, https://www.osha.gov/hydrogen-sulfide/hazards

US Chemical Safety and Hazard Investigation Board (CSB), Aghorn Operating Waterflood Station Hydrogen Sulfide Release, https://www.csb.gov/aghorn-operating-waterflood-station-hydrogen-sulfide-release-/

Click 2 Houston News – October 11, 2024, PEMEX confirms 2 dead, 13 hospitalized after deadly gas release at Deer Park facility, https://www.click2houston.com/news/local/2024/10/11/pemex-confirms-2-dead-13-hospitalized-after-deadly-gas-release-at-deer-park-facility/

Frontline Data Solutions, Hydrogen sulfide in the oil and gas industry, https://www.fldata.com/hydrogen-sulfide-oil-gas-industry

US Occupational Safety and Health Administration (OSHA), Hydrogen Sulfide, https://www.osha.gov/hydrogen-sulfide

About the Author

M. Darryl Yoes is an experienced refinery supervisor and manager, having spent over 50 years in the refining industry. He began his career as a Process Apprentice at one of the largest US Gulf Coast refineries. After quickly advancing through several supervisory and managerial assignments, including first- and second-line supervisor positions in Process, Mechanical, and Technical, he advanced to the position of Risk Management Advisor. He then became the Safety, Health, and Environmental Manager of one of the largest petroleum refineries in the world.

He has led or participated in numerous process safety incident investigations, including those involving the tragic loss of life. He served as an emergency coordinator and incident manager in the role of Refinery Superintendent and has extensive experience managing other emergency operations, including refinery fires, explosions, and oil spill response operations. He spent two years as a group leader on the Valdez, Alaska Spill Response Team. He has also led or participated in numerous process hazards analysis studies and led or participated in operations integrity and safety risk assessments at major US, European, and Asian petrochemical sites. He has also served as the Safety Manager for major construction projects at US refineries.

Mr. Yoes continues process safety management consulting for refining and petrochemical plants and for construction safety management for major construction projects across the United States and is a course leader and facilitator for petrochemical plant Operations Safety Management Training designed for refinery and petrochemical plant management, supervision, engineering, and plant operators. He has led comprehensive Process Safety Operations Training courses in essentially all regions of the world, including North and South America, Asia, Australia, and Europe, to share lessons learned from industry process safety incidents involving loss of containment of flammable or toxic materials leading to fire, explosions, and injuries or fatalities. The Safe Operations Training presents practical methods or work practices to help avoid a recurrence of these types of incidents.

He is the author of two other Wiley textbooks and online process safety resources. The first textbook is titled *Process Operations Safety – The What, Why, and How Behind Safe Petrochemical Plant Operations*, which is available at most bookstores, Amazon, and online. This book is a comprehensive overview of process safety in petroleum refineries and petrochemical plants. It is a valuable resource to operators, plant supervisors, engineers, managers, and others responsible for process operations safety. To a significant extent, the content of the Darryl Yoes Safe Operations Training course is reflected in this book. The second book, titled *Hazards of Nitrogen and Other Inert Gases: How They Can Be Safely Managed*, was recently released for purchase, also at Wiley online, Amazon, and your neighborhood bookstore.

Mr. Yoes is also the founder and president of Safety Consulting International, LLC, which developed these helpful guides to help prevent further process safety incidents. He is employed by EcoScience Resource Group, LLC, in Baton Rouge, a firm specializing in developing process operations training manuals, process operating and environmental procedures, and operator training guides for the petroleum refining and petrochemical industries.

1

The Sources of Hydrogen Sulfide: Where Does It Come From?

Hydrogen sulfide (H_2S) is readily found in many places worldwide. Many of these are due to natural processes, such as decaying organic matter, animal or human waste, decaying vegetation, and natural volcanic or geothermal formations. This is especially true of hot springs, which also contain sulfur. We recently visited Yellowstone National Park and couldn't help but notice the rotten egg odor near the hot springs. It is common to find H_2S in swamps or bogs with abundant decaying vegetation that produces hydrogen sulfide. This natural formation of H_2S is a process referred to as anaerobic digestion. This occurs due to the bacterial breakdown of organic matter, such as decaying vegetation or animal waste, in oxygen-deprived (anaerobic) conditions.

Hydrogen sulfide is usually plentiful in sewer systems or treatment facilities due to the naturally decaying organic waste. It is also commonly found in landfills or other waste treatment facilities, especially in the leachate (trapped or draining liquids) from these facilities. Animal farms, ranches, or dairies are challenged to manage animal waste due to the potential formation of hydrogen sulfide in the waste management drainage, trenches, or storage facilities. For example, manure pits are notorious for accumulating hydrogen sulfide and must be managed accordingly.

In summer, hot weather typically significantly accelerates the decay of most organic materials, such as manure and vegetation, increasing the generation of hydrogen sulfide. This is especially true for confined spaces or areas with limited air circulation. Higher temperatures increase the generation of H_2S, and low air circulation results in higher concentrations in confined spaces.

So, how do these exposures happen? Think about a dairy farmer, cattle rancher, or pig farmer and how much animal waste they must handle daily. Many of these occupations dig pits or have storage tanks or pits where waste is stored for a period, and then, they haul the waste to an off-site storage site or a processing site for treatment or disposal. This has developed into a big business since some states

Hydrogen Sulfide (H_2S) Hazards: Risk Recognition and Prevention, First Edition. M. Darryl Yoes.

provide financial incentives to cattlemen and dairy farmers to transform manure into a valuable natural gas product. The waste is fed to anaerobic digesters, which convert it into methane or natural gas in a low-oxygen environment and into the natural gas pipeline network.

However, incidents have occurred when they, or one of their helpers, enter the waste pit or storage tank for some reason. A common occurrence of this type is retrieving a dropped tool or something similar. When they enter the confined space, they are unexpectedly exposed to the toxic vapors of hydrogen sulfide and are killed.

Another whole series of cases has occurred at municipal sewer treatment plants where workers enter underground pits to perform work on waste transfer pumps or underground piping and are unexpectedly exposed to the H_2S vapors and are killed. Unfortunately, in these cases, there is generally more than one fatality involved, as other workers, driven by the natural human instinct to rescue their friend or coworker, also rush in, are exposed, and die from the toxic nature of H_2S. This and other tragic incidents due to H_2S exposures are covered in much more detail in this textbook and Appendix 2.

Oil and Gas Formations and Production Wells: A Significant Source of Hydrogen Sulfide

Hydrogen sulfide (H_2S), sometimes called "sour gas," occurs naturally in most oil and gas wells. It is present in the formations where these wells are drilled, sometimes hundreds of feet below the surface, and can be tens of thousands of feet below the surface. Geologists and petroleum geologists believe these formations were formed more than 500 million years ago by the natural degradation of dead organisms from plants and animals, like phytoplankton (plants) and zooplankton (animals). This dead organic matter formed layers and, over millions of years, was covered by the sea or earth, resulting in exposure to both pressure and heat and forming underground seas of hydrocarbons.

Bacterial action most likely resulted in the formation of hydrogen sulfide through a process called microbial sulfate reduction (MSR). MSR is a process where the sulfate-containing minerals are converted to H_2S by bacteria in the presence of water. Another process, thermochemical sulfate reduction, is one that occurs at high temperatures and where no bacteria are required. These processes occur deep within the earth where the oil and gas are trapped under heat and pressure, resulting in H_2S in the crude oil and gas formations. Since oil and gas were formed by the natural degradation of dead plants and animals, we often refer to these products as "fossil fuels." These are molecules of carbon and hydrogen bonded together to form the hydrocarbons, which we process into various fuels and thousands of other products.

Oil and gas production wells are significant sources of hydrogen sulfide. Along with the valuable hydrocarbons, the gas and oil, the wells also contain significant amounts of hydrogen sulfide and many other sulfur compounds, which are contaminants and must be removed from the oil and gas, either in the gas fields or in the petroleum refineries while processing the crude oil. Some production wells have sulfur removal capability, with a smaller process unit on-site designed to remove hydrogen sulfide from the produced gas and convert it into elemental sulfur, which is sold as a product.

Produced water is also a product from oil and gas wells and is very high in hydrogen sulfide content and other sulfur compounds. This sour water is typically produced to storage tanks near the wellhead and later injected back into the formation to maintain well pressure and serve as a disposal method for the sour water. The H_2S content of this sour water represents a hazard to the operators, and they must be aware of these hazards and wear personal protective equipment when working on or near this equipment.

A significant amount of sulfur compounds and hydrogen sulfide from the oil and gas wells come into the refinery in the crude oil and petroleum distillates and must be removed during the refining process to avoid going to the consumer in the refined products. Intermediate products are further refined by hydrotreating to remove the sulfur compounds before they go into finished products like gasoline, diesel, lubricants, and a wide range of petroleum products. This is done in the refinery, and this process is described in the next few paragraphs.

Petroleum Refining: A Significant Source of Hydrogen Sulfide

In a typical petroleum refinery, hydrogen sulfide and many other sulfur compounds are very corrosive to refining equipment and must be removed early in the refining process. Also, if allowed in the products produced by the refinery, such as gasoline, diesel, or lubricating oils, they can adversely affect the products, resulting in fouling or corrosion to the vehicles or other equipment. The most common process for removing the sulfur compounds from the intermediate refining products is scrubbing with amines. Amine treatment is typically an integral part of hydrotreating, hydrocracking, or similar hydroprocessing operations. Several other processes are available to treat refining gases to remove sulfur compounds. These include contact with absorbents, treatment with hydrogen peroxide, another product called Pro3® GT or Pro3® GT+, and others. However, the primary treatment method for H_2S removal in petroleum refineries is having the gas or product in circulation contact with an amine.

Typically, the hydroprocessing units (hydrotreaters, sometimes known as hydrofiners or hydrodesulfurization [HDS] units) and hydrocrackers convert the sulfur compounds that come in with the feed into hydrogen sulfide in a reaction involving high temperatures, hydrogen, and pressure in the presence of a fixed catalyst bed. The catalytic reaction converts the sulfur compounds to hydrogen sulfide. The hydroprocessing products, typically sour naphtha containing H_2S, are then scrubbed with amines before further processing in the refinery. The amines have an affinity for hydrogen sulfide, and the H_2S is absorbed by the amine and carried away in the foul amine (sometimes referred to as rich amine or amine-containing H_2S). The foul amine is then regenerated in a separate vessel by stripping the amine with temperature and steam. This strips the hydrogen sulfide from the amine, allowing the amine to be recovered and reused in the process as lean amine. The super-concentrated hydrogen sulfide vapors, typically 80–90% H_2S (800,000–900,000 ppm), are then sent to the sulfur recovery unit, where the H_2S is converted to elemental sulfur for sale as a separate product. This process removes essentially all the H_2S and other sulfur compounds from the final products, and after further processing and blending, the products, typically gasoline, diesel, and lubricating oils, are blended to meet all other product specifications before being sent to the consumer.

Figure 1.1 illustrates the location and function of an amine absorber and amine stripper and how they are integrated into a typical naphtha hydrotreater unit. The hydrotreater products flow from the cold separator, which is downstream of the hydrotreater or hydrocracker reactor and is routed through the amine absorber in counterflow with the circulating amine, where the sulfur compounds, primarily hydrogen sulfide, are absorbed into the amine.

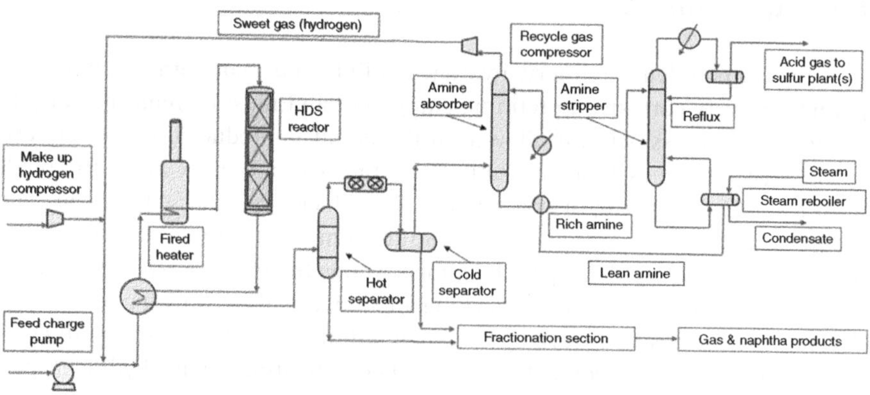

Figure 1.1 The location and function of an amine absorber and amine stripper on a typical naphtha hydrotreater unit.

The H_2S is then stripped from the amine in the amine stripper, and the lean amine is circulated back to the amine absorber to continue the operations. The acid gas, which is super-rich in hydrogen sulfide, flows from the top of the amine stripper and is routed to the sulfur plants, where it is converted into elemental sulfur for sale. Acid gas is the stream that may be as high as 80–90%, or even higher, in hydrogen sulfide.

The lean amine from the bottom of the amine stripper flows to the top of the amine absorber, where it contacts the circulating process vapors and liquids, absorbing the hydrogen sulfide and other sulfur compounds. The "sweet gas," which is the hydrocarbon product after scrubbing, leaves the top of the amine absorber and goes back into the process as the sweet product.

The rich amine circulates back from the bottom of the amine absorber to the top of the amine stripper, where the H_2S is stripped with heat and steam. The super-rich H_2S, or acid gas, flows from the stripper overhead to the sulfur plants, where it is converted to elemental sulfur.

Even though the technology is similar, there are several configurations for amine treatment in a petroleum refinery. For example, some large multi-train refineries may have amine absorbers located on each of several different hydrotreaters treating naphtha, diesel, reformer feed, and other similar process streams. In this configuration, the amine stripper(s) may be located near the sulfur plants to minimize the distance the acid gas containing super-rich concentrations of H_2S must travel. In this configuration, they may have a piping distribution system for lean amine and rich amine circulation between the amine absorbers and strippers. The acid gas pipelines may contain super-rich H_2S concentrations, 80–90% or higher (800,000–900,000 ppm).

Additional Discussion on Amine Absorber/Stripper Operations

Figure 1.2 shows a simplified example of a consolidated absorber/stripper for a single naphtha hydrotreater, where the two towers are located on one process unit and are typically in proximity, one to the other. This is what I would expect to see in a smaller, single-train refinery.

In this example, the sour gas from the hydrotreater hot separator enters the amine absorber near the bottom of the absorber tower. It travels in countercurrent flow with the downflowing lean amine (typically monoethanolamine [MEA], diethanolamine [DEA], or methyl diethanolamine [MDEA]). The tower is designed to ensure maximum contact of the amine solution with sour gas, allowing the amine to absorb the hydrogen sulfide into the amine solution.

The amine solution containing the H_2S, now referred to as rich amine or sometimes called foul amine, leaves the bottom of the absorber tower to be stripped of

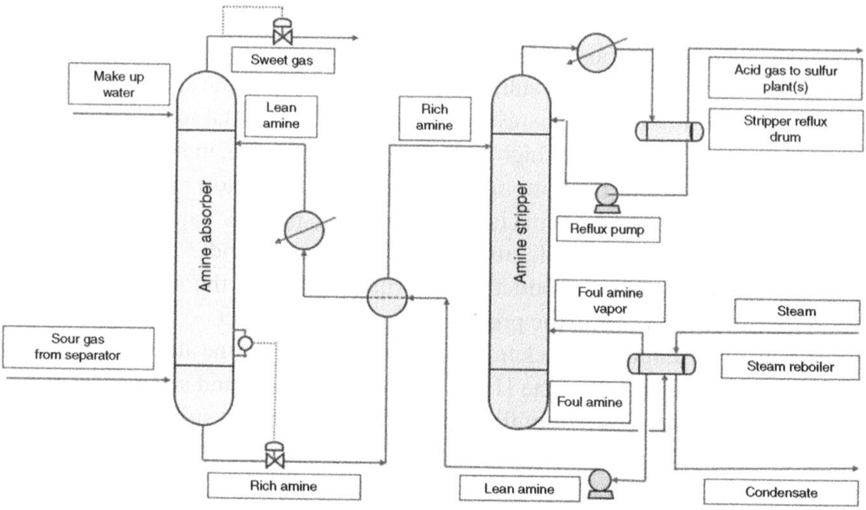

Figure 1.2 A close-up view of the integration of an amine absorber and amine stripper on a typical naphtha-integrated hydrotreater unit.

the H_2S in the amine stripper. The absorber overhead releases the scrubbed gas, now containing very little H_2S, and referred to as sweet gas, into the process gas system for further processing.

The rich amine, sometimes referred to as foul amine, containing a very high concentration of H_2S, goes through the amine exchanger, where it is preheated by exchanging heat with the lean amine from the amine stripper bottoms. The rich amine then enters the amine stripper near the top of the tower and travels down the tower, contacting the hot amine vapors from the steam reboiler. The hot vapor strips the H_2S from the foul amine, and the H_2S exits the stripper as super-high concentration H_2S, where it goes to the sulfur plants for conversion to elemental sulfur for sale. With very little H_2S, the lean amine returns to the amine absorber to continue the process.

Inadvertent Mixing of Chemicals: Another Source of Hydrogen Sulfide

Another way we can inadvertently generate hydrogen sulfide is through the unintentional mixing of two or more incompatible chemicals. For example, the offloading of a truck carrying a process chemical into a tank containing a different and incompatible chemical. This has happened and has resulted in the generation of hydrogen sulfide or another toxic chemical.

Another example is the inadvertent mixing of acid with sour water (water containing hydrogen sulfide). This will result in a violent chemical reaction and instantaneously release copious amounts of hydrogen sulfide into the atmosphere. Incidents of this type have occurred during the mixing of two different chemicals where the mixture was not adequately controlled, resulting in the generation of H_2S, which has resulted in fatalities.

The inadvertent mixing of most sulfides with an acid will generate hydrogen sulfide (H_2S). For example, the inadvertent mixing of a chemical solution of sulfides with hydrochloric acid, sulfuric acid, or almost any acid generates a large release of hydrogen sulfide gas (H_2S). When this happens, it is usually due to unloading one of the chemicals into the wrong storage tank; for example, offloading a tank truck of sodium hydrosulfide into a tank of sulfuric acid or mixing iron sulfides with an acid. This has occurred, resulting in fatalities and large evacuations of the surrounding community.

Unloading a truckload of almost any compound with the name sulfide into a tank containing a dilute acid may release copious amounts of hydrogen sulfide gas. There have been incidents like this when a chemical is unloaded into the wrong receiving tank, inadvertently mixing the two chemicals. Dangerous chemicals like hydrogen sulfide and chlorine have been released when this happens.

This has happened, resulting in the death of the truck driver and a large-scale evacuation of the surrounding community. A comprehensive publication from Wiley covers the expected result of mixing two incompatible chemicals titled *Wiley Guide to Chemical Incompatibilities, 2nd Edition*. Another Notre Dame University article lists chemicals and their incompatibility with other chemicals. Notably, these highlight sulfides as being incompatible with acids. Please refer to the Additional Resources section of this chapter for links to both publications. This is discussed more in the case studies in "What Has Happened/What Can Happen Again."

Decaying Organic Materials such as Landfills, Sewerage, Animal Carcasses, Grain, or Other Vegetation (or Other Animal Matters such as Blood, Chicken Parts, Milk)

Hydrogen sulfide is a natural byproduct when organic material is allowed to decay, especially in a low-oxygen (anaerobic) environment. Hydrogen sulfide is produced naturally when bacteria break down the dead plant and animal material, causing it to decay resulting in an off-gas that is flammable and has a foul "rotten egg" odor. If organic material such as grain, corn, or other plant or animal matter is left in a pit in the ground, a storage silo, or a tank truck, especially where it becomes anaerobic (little or no oxygen), the decay process can generate significant concentrations of hydrogen sulfide. This is more likely to happen when the organic material is covered by stagnant water or buried in mud, where it is likely to become anaerobic.

Anaerobic conditions most often occur in bogs or swamps where the water is stagnant, but can also happen in tanks, silos, or pits. The bacteria, especially the sulfate-reducing bacteria, can convert sulfur compounds within the decaying organic matter into H_2S. The hydrogen sulfide can be deadly when a cattle rancher or farmer enters a pit to unplug a drain, or an unsuspecting worker enters a confined space that contains decaying grain, or a truck driver opens a truck compartment containing decaying animal matter and is overcome by the deadly vapors (H_2S).

Examples of Industrial Processes with Potential for Hydrogen Sulfide (This Is Not Intended to Be a Complete List)

Hydrogen sulfide is also either used by the following or may be a byproduct in these industrial processes:

- Petroleum production (oil and gas wells)
- Petroleum refining
- Some petrochemical facilities
- Operations that create sulfuric acid (H_2S and sulfur are feedstocks for these operations)
- Municipal sewer treatment plants
- Agricultural grain silos
- Textile manufacturing
- Pulp and paper processing (paper mills)
- Food processing
- Hot asphalt paving
- Mining
- Tanning operations
- Ore refineries
- Rayon manufacturing
- Coal gasification plants
- Intensive livestock operations

Chapter 1: Key Lessons Learned

Hydrogen sulfide is naturally occurring and is found in many places worldwide, especially where natural biological degradation processes are occurring, such as decaying vegetation or human and animal waste. It is also found in naturally occurring volcanic activity and in some water supplies due to decaying vegetation in underground deposits. Along with other sulfur compounds, hydrogen sulfide is also naturally occurring in crude oil and gas wells.

Many sulfur compounds are converted to hydrogen sulfide in the petroleum extraction and refining processes and other manufacturing processes such as pulp and paper processing, ore refining, and coal gasification plants. Amine chemicals, typically MEA, DEA, or MDEA, are used in petroleum refining operations. These chemicals accumulate super-concentrated hydrogen sulfide when treating refining products to remove sulfur compounds. The super-rich hydrogen sulfide gas is sent to the sulfur unit where, with temperature and catalyst, it is converted to elemental sulfur or further processed to produce sulfuric acid.

Petroleum refineries have streams in processing facilities that are super-rich with hydrogen sulfide. These streams may have hydrogen sulfide concentrations as high as 80–90% or higher (800,000–900,000 ppm). This requires special procedures and practices for any work on or near this equipment to prevent a loss of containment of the highly toxic, high-concentration hydrogen sulfide.

Hydrogen sulfide is also found in sewerage treatment plants, underground sewer and wastewater piping, and pump transfer operations. Large cattle ranches and dairy farms may accumulate animal waste in temporary pits or storage tanks, and typically, the waste generates significant accumulations of hydrogen sulfide in the vapor space above the accumulated waste.

Hydrogen sulfide can also be generated from the inadvertent mixing of incompatible chemicals, such as sour water (water containing sulfur compounds) and acid, iron sulfide with diluted acid, and other sulfide-containing chemicals with acid solutions. This has occurred because the wrong chemical was transferred into an on-site storage tank, resulting in the inadvertent mixing of incompatible chemicals. Large releases of highly toxic hydrogen sulfide gas have been generated this way.

Chapter 1: Incident(s) Related to the Key Lessons Learned

Hydrogen Sulfide Incompatibility with Acids

I have some personal experience with the inadvertent mixing of chemicals. One evening at our refinery, we had a sour water stream flowing into a process sewer due to an unexpected failure and outage of a sour water pump. This sour water was safely and temporarily processed in the wastewater treatment plant. However, at about the same time, another unit had an unexpected release of sulfuric acid into the same process sewer, resulting in a mixture of the sour water containing H_2S with the sulfuric acid in the sewer system. This unplanned and unexpected mixing of two incompatible chemicals resulted in a reaction between the acid and sour water, which released a significant amount of H_2S vapor, which began fuming from the process sewers. As you might guess, this was unexpected and, for a few minutes, presented a serious situation.

This was a long time ago, but this incident is a good reminder of the hazards of mixing chemicals that may be incompatible. This has happened in the industry in other ways. These include unloading a shipment of chemicals into the wrong tank or the inadvertent mixing of incompatible commercially available cleaning chemicals. We must have procedures and practices to ensure that we verify the shipping documents from the chemical supplier and the labels on the tank and unload connections for every shipment to ensure they match. It is also important to verify the safety data sheet before mixing chemicals to ensure compatibility.

Additional Resources

The Occupational Safety and Health Administration (OSHA), OSHA Fact Sheet Hydrogen Sulfide (H_2S), https://www.osha.gov/sites/default/files/publications/hydrogen_sulfide_fact.pdf

Agency for Toxic Substances and Disease Registry, Office of Innovation and Analytics, Hydrogen Sulfide – ToxFAQs™, CAS # 7783-06-4, https://www.atsdr.cdc.gov/toxfaqs/tfacts114.pdf

Agency for Toxic Substances and Disease Registry, PUBLIC HEALTH STATEMENT Hydrogen Sulfide, https://www.atsdr.cdc.gov/toxprofiles/tp114-c1-b.pdf

The National Library of Medicine; National Center for Biotechnology Information, PubChem – Hydrogen Sulfide, https://pubchem.ncbi.nlm.nih.gov/compound/Hydrogen-Sulfide

The Essential Chemical Industry – Online, Sulfuric Acid, https://www.essentialchemicalindustry.org/chemicals/sulfuric-acid.html

Sulfur Recovery Engineering Inc, The Claus Sulfur Recovery Process, https://www.sulfurrecovery.com/sulfur-recovery-process

The National Fire Protection Association (NFPA), The Hazard of Mixing Chemicals, https://www.nfpa.org/news-blogs-and-articles/blogs/2023/02/20/the-hazard-of-mixing-chemicals

Wiley Guide to Chemical Incompatibilities, 2nd Edition, Richard P. Pohanish, Stanley A. Greene, https://www.wiley.com/en-us/Wiley+Guide+to+Chemical+Incompatibilities%2C+2nd+Edition-p-9780471721628

The University of Notre Dame, Incompatible Chemicals, https://riskmanagement.nd.edu/assets/444930/incompatible_chemicals.pdf

The Texas Tribune, Study finds levels of a dangerous gas "off the scales" in Central Texas oilfield, https://www.texastribune.org/2024/09/20/texas-oil-wells-hydrogen-sulfide-caldwell-county/

Chapter 1: End-of-Chapter Quiz

1. Where are we likely to find significant amounts of hydrogen sulfide?
   Please choose all that apply.
   a. Swamps or bogs with rotting vegetation
   b. Oil and gas well condensate
   c. Liquids draining from waste disposal sites
   d. Sour water drained from crude oil storage tanks
   e. Agricultural grain silos with rotting corn or grain
   f. None of the above
   g. All of the above

2. In which industries are we most likely to find significant concentrations of hydrogen sulfide?
   Please choose all that apply.
   a. Leather tanning
   b. Petroleum refining
   c. Wastewater treatment
   d. Pulp and paper mills
   e. Textile manufacturing
   f. Mining
   g. None of the above
   h. All of the above

3. When a sour water stream, water containing hydrogen sulfide, contacts ______________, a significant amount of hydrogen sulfide gas may be released into the atmosphere.

4. Most petroleum refineries process hydrogen sulfide in concentrations as high as __________ to __________ percent.

5. Major H_2S releases have occurred by the accidental ______________ of a truckload of incompatible chemicals into the wrong tank.

6. The most common way H_2S is removed from the product in a petroleum refinery is by scrubbing the product with an ____________ chemical.

7. Please describe the purpose and function of a hydrotreater unit in a petroleum refinery.
   Please select the most correct answer.
   a. The hydrotreater removes sulfur compounds from the crude oil before it goes to the crude units.
   b. The hydrotreater ensures that the product meets consumer specifications.

c. The hydrotreater removes sulfur compounds from the naphtha product using a fixed catalyst, hydrogen, and heat.
d. The hydrotreater is the most important unit because it produces clean hydrogen.
e. The hydrotreater.

8. What function does the amine absorber play in a petroleum refinery hydrotreater unit?
Please select the most correct answer.
a. The amine absorber sends the foul amine to the sulfur plant.
b. The amine absorber removes the hydrogen sulfide from the gas streams by absorbing it onto amine.
c. The amine absorber cleans the foul amine.
d. The amine absorber is in the feed system to the hydrotreater and purifies the feed to the reactor.

9. What is likely to happen if sour water (water containing H_2S) is mixed with any form of acid?

10. Hydrogen sulfide is also found in ______________ treatment plants and underground sewer and wastewater piping and pump transfer operations. Also, large __________ ranches and __________ farms may accumulate animal waste in temporary pits or storage tanks, and typically the waste will generate significant accumulations of hydrogen sulfide in the vapor space ________ the accumulated waste.

11. More typically in summer, the hot weather significantly ______________ the decay of most organic materials, such as manure and vegetation. This ______________ the hydrogen sulfide generation.

2

How Biodegradation Processes Generate Hydrogen Sulfide

Hydrogen sulfide (H_2S) is most commonly obtained by its separation from sour gas, natural gas, or other petroleum hydrocarbons containing a high hydrogen sulfide concentration. Hydrogen sulfide is also produced by contacting hydrogen with elemental sulfur at an extremely high temperature, typically above 840 °F (450 °C). Sometimes, hydrocarbons are the source of hydrogen when used in this process.

At wastewater treatment facilities, hydrogen sulfide can be produced from the microbial breakdown of organic matter in the absence of oxygen. This process is commonly known as anaerobic digestion and occurs in equipment referred to as anaerobic digesters. It is accomplished by treating the wastewater with sulfate-reducing microorganisms under anaerobic conditions.

However, it is also extremely important for dairy farmers, cattlemen, and ranchers to understand that hydrogen sulfide (H_2S) can be readily formed from the natural decay processes, specifically the biodegradation of cattle or horse manure. This occurs when sulfate-reducing bacteria (SRB) break down organic matter that contains sulfur compounds, such as proteins. Sulfur must be present for H_2S to form, and the source of sulfur comes from sulfur-containing amino acids in organic matter, mainly cysteine and methionine. This is especially true when biodegradation occurs in an environment without oxygen, otherwise known as an anaerobic condition. Under anaerobic environmental conditions, in the presence of chemical oxygen demand, the SRB can convert sulfate in the wastewater to sulfide, which then combines with hydrogen to produce hydrogen sulfide gas (H_2S). This process occurs through the decomposition of sulfur-containing compounds at the microbial level in organic matter like cysteine and methionine, which are typically found in proteins. Cysteine and methionine are both sulfur-containing amino acids found in proteins. When the microbes break them down, the sulfur compounds are converted to hydrogen sulfide gas and released into

Hydrogen Sulfide (H_2S) Hazards: Risk Recognition and Prevention, First Edition. M. Darryl Yoes.
 Published 2026 by John Wiley & Sons, Inc.

the environment. SRB play a crucial role by converting sulfate (SO_4^{2-}) into sulfide (S^{2-}), which then reacts with hydrogen to form H_2S.

Microbes, when at work in the manure, especially in an anaerobic environment, break the manure down into an inorganic portion, the largest fraction of the matter, and the organic portion, also known as the humus portion. The inorganic portion, the largest fraction, is primarily made of decomposed rocks broken down into sand, loam, and clay. The organic portion contains the decomposed protein and other organic matter formed by the anaerobic biodegradation process of the manure. As a reminder, hydrogen sulfide is primarily formed when biodegradation occurs in an anaerobic environment. This means most H_2S production due to biodegradation happens in environments with little or no oxygen, such as swamps, sewers, wastewater treatment plants, and cattle manure storage tanks or bins. Hydrogen sulfide can and has also occurred deep within landfills, where bacteria can utilize sulfate as an electron acceptor.

The following is a little more information on the forces of decomposition at work when organic matter like sewage decomposes, resulting in hydrogen sulfide generation. As stated earlier, H_2S is formed when the SRB are at work, especially in the absence of oxygen or what we refer to as anaerobic conditions, converting sulfates into sulfides. The sulfides then bond with hydrogen molecules from contact with the wastewater, forming the toxic hydrogen sulfide gas. Should you desire a more technical discussion, please see the links in the Additional Resources section at the end of this chapter.

When sewage is decomposing, a process known as microbial metabolism breaks the organic matter such as proteins, fats, or carbohydrates into simple compounds such as carbon dioxide, water, and biomass, or organic material from living or recently living organisms, such as plants and animals. This can occur in an environment where oxygen is available to help drive microbial metabolism. Some wastewater treatment plants use this type of process to treat sewer wastewater. Generally, no toxic gases like hydrogen sulfide are released under these aerobic conditions.

However, when little oxygen is available and the conditions are anaerobic, the SRBs actively consume sulfate, generating sulfide or the base chemical for hydrogen sulfide. The SRBs use sulfate (SO_4^{2-}) as an electron acceptor during their metabolism, especially in the absence of oxygen. SRBs thrive in anaerobic conditions, whereas many other organisms, including humans, cannot survive.

Areas that are particularly susceptible to the development of hydrogen sulfide are the layers of slime or gelatinous matter that can accumulate on internal pipe walls in underground sewer piping and in sewer processing equipment such as septic tanks, sludge pits, sewer pumping plants, and the sludge blankets of wastewater clarification units and other solids processing units. When the dissolved oxygen is depleted, the sewer quickly becomes anaerobic; the sewage becomes septic. This is when the SRBs thrive and quickly reduce the sulfur compounds or sulfates to hydrogen sulfide. Hydrogen sulfide is highly soluble in wastewater and can generate a weak acid called hydrosulfuric acid, which is highly corrosive to concrete piping and some metals. Hydrosulfuric acid also has the formula H_2S but is a liquid form as a weak acid.

The rate of hydrogen sulfide development in sewer systems is accelerated by the following process conditions: low flow (velocity) of sewage in piping or in the accumulated slime waste, resulting in longer retention time, a piping system that is nearly horizontal with little or no downward slope, an internal piping system that remains wetted with waste material; environmental conditions that result in the temperature remaining warm; low availability of oxygen; and the presence of sulfates in the waste material.

Chapter 2: Key Lessons Learned

- Hydrogen sulfide (H_2S) is most commonly obtained by its separation from sour gas, natural gas, or other petroleum hydrocarbons containing a high concentration of sulfur compounds, including hydrogen sulfide.
- Hydrogen sulfide is also prone to form in areas with a lack of available oxygen, otherwise known as anaerobic conditions. These can occur when manure or other organic matter is decaying in pits, bogs, or at the bottom of tanks or silos. The SRBs react with the organic matter and convert sulfate into sulfide, which then reacts with hydrogen to form H_2S.
- Significant amounts of hydrogen sulfide can be released when the decaying matter is stirred or disturbed and can result in the deaths of workers or others who may be nearby.

Chapter 2: Incidents Related to Key Lessons Learned

August 2021 – Miami Herald News Article: 3 Fatalities (3 Brothers)

Three brothers die after passing out in a manure pit at an Ohio farm, authorities said, by Mitchell Willetts, August 11, 2021, 12:59 p.m. At the time of the Tuesday afternoon accident, the brothers – Gary, Todd, and Brad Wuebker – were performing maintenance on a pump inside the storage pit, the *Mercer County Outlook* reported. Manure pits are known to produce a wide range of toxic gases. St. Henry firefighters received a call for help around 12:30 p.m., the paper reported, and when they arrived at the livestock operation, all three men were unconscious in the pit.

August 2002 – OSHA Database: 1 Fatality (Attempted Rescue)

On August 27, 2002, Employee #1 and a coworker were installing a new gate and removing the old one from a stand box filled with manure water from a manure pond. The dairy owner ran the pump to draw water out of the box and reduce the water level below the opening port of the gate. When the water level had been lowered, Employee #1 entered the box, removed the old gate, and cleaned the old sealing mortar from around the opening. He and the coworker then lowered the new gate into the box. Employee #1 reentered the box to position the gate, while the coworker and one of the dairy owners stood outside the box and held onto the gate from above. They were watching Employee #1 when he suddenly rolled onto his right side and landed, back down, in the water, with his head submerged. The coworker entered the box to rescue Employee #1, passed out momentarily, regained consciousness, and then climbed out of the box under his own power. Emergency responders were called, and Employee #1 was extricated from the stand box. He was transported to the hospital, where he was pronounced dead from the inhalation of hydrogen sulfide and possibly methane. Hydrogen sulfide levels were later measured at 218 PPM under conditions approximating those of the accident.

Additional Resources

Science Direct, Electron Acceptor, https://www.sciencedirect.com/topics/chemistry/electron-acceptor

Science Direct, Sulfate-Reducing Bacteria, https://www.sciencedirect.com/topics/agricultural-and-biological-sciences/sulfate-reducing-bacteria

National Library of Medicine, National Center for Biotechnology Information, Domenick J Braccia, Xiaofang Jiang, Mihai Pop, A Brantley Hall, https://pmc.ncbi.nlm.nih.gov/articles/PMC8564485/#:~:text=There%20are%20two%20major%20sources,et%20al.%2C%202012)

Wikipedia, Hydrogen Sulfide, https://en.wikipedia.org/wiki/Hydrogen_sulfide#:~:text=Hydrogen%20sulfide%20is%20often%20produced,done%20by%20sulfate%2Dreducing%20microorganisms

National Library of Medicine, The Center for Biotechnology Information, Methionine Oxidation and Reduction in Proteins, https://pmc.ncbi.nlm.nih.gov/articles/PMC3766491/

The Impact of Wastewater Quality and Flow Characteristics on H2S Emissions Generation: Statistical Correlations and an Artificial Neural Network Model, Mohsina Sherief, Ashraf Aly Hassan, https://www.mdpi.com/2073-4441/14/5/791

Chapter 2: End-of-Chapter Quiz

1. At wastewater treatment facilities, hydrogen sulfide can be produced from the ________ breakdown of organic matter in the absence of __________.

2. At wastewater treatment plants, the breakdown of organic matter is referred to as ________ __________ and occurs in equipment referred to as anaerobic digesters.

3. It is also extremely important for dairy farmers, cattlemen, and ranchers to understand that hydrogen sulfide (H_2S) can be readily formed from the natural decay processes, or the ____________ process in cattle or horse manure.

4. This occurs when bacteria, especially ________-__________ (SRB), break down organic matter containing sulfur compounds, such as proteins.

5. The sulfate-reducing bacteria (SRB) can convert sulfate in the wastewater to sulfide, which then combines with __________ to produce hydrogen sulfide gas (H_2S).

6. Hydrogen sulfide is primarily formed when the biodegradation occurs in an __________ environment. This means that most H_2S production due to biodegradation happens in environments with little or no __________.

7. ______'s are known to play a crucial role in the process by converting sulfate into sulfide, which then reacts with hydrogen to form H_2S.

8. After the bacteria break down the organic matter, the largest portion is primarily sand, loam, and clay, while the organic fraction is known as ________.

9. Significant amounts of hydrogen sulfide can be released when the decaying matter is stirred or __________ and can result in the __________ of workers or others who may be nearby.

10. Hydrogen sulfide can also occur deep within landfills where there is a lack of ________.

3

The Characteristics and Effects of Hydrogen Sulfide

Toxicity

Hydrogen sulfide (H_2S) is typically a colorless gas with a very distinct rotten egg odor. This odor is very noticeable in low concentrations. It is normally easily detectable in concentrations as low as 0.1–1.5 parts per million (ppm), and the odor becomes more offensive at around 3–5 ppm.

However, it is a "devious" and "deceitful" gas because at concentrations of 100–150 ppm, it paralyzes the olfactory nerves, and the victim can no longer smell it. This makes the victim believe that the gas has dissipated and is no longer a threat until it is too late. This characteristic makes H_2S a very dangerous gas and is one reason why there are so many deaths involving exposure to H_2S. The sense of smell should never be used to identify or locate H_2S.

The difference between H_2S being a nuisance and becoming deadly can be very quick. For example, at about 50 ppm, it is certainly an offensive nuisance, but it is not deadly. The sense of smell disappears at 100–150 ppm, and at 300–500 ppm, death can occur after several hours. However, at 1,000–2,000 ppm, knockdown is immediate, and death is almost instantaneous.

Continued exposure in low concentrations can result in nasal congestion and respiratory issues, including coughing, shortness of breath, and chest pain. These exposures can result in eye irritation, burning in the eyes and nose, and as stated previously, the loss of sense of smell. Also, exposure to low concentrations can result in headaches, dizziness, and confusion. The victim becomes confused and may have difficulty understanding what is happening and how to escape. Continued exposure to low concentrations can also result in nausea and vomiting.

At higher concentrations of about 100–150 ppm, H_2S will quickly deaden the sense of smell to the point where the victim believes the gas has dissipated and is

Hydrogen Sulfide (H_2S) Hazards: Risk Recognition and Prevention, First Edition. M. Darryl Yoes.

no longer a threat. This can happen almost instantaneously in these concentrations, making them unaware of the still-present hazard.

Continued work in these concentrations can result in neurological effects, including memory loss, poor attention span, and reduced motor function (ability to move arms and legs, etc.). This can also result in chronic respiratory problems, including bronchitis and asthma, eye and skin damage, leading to chronic irritation, redness, and rashes, and chronic headaches, including debilitating migraines.

As the concentration or duration of exposures increases, so do the symptoms and effects. This is illustrated in Table 3.1, which is provided courtesy of the Occupational Safety and Health Administration (OSHA). Please note that the loss of smell occurs at 100–150 ppm. At 500–700 ppm, the victim may be staggering and collapse after 5 minutes, and exposure over 700 ppm can result in collapse, unconsciousness, and

Table 3.1 Selected physical and chemical properties.

Concentration (ppm)	Symptoms/Effects
0.00011–0.00033	Typical background concentrations
0.01–1.5	Odor threshold (when rotten egg smell is first noticeable to some). Odor becomes more offensive at 3–5 ppm. Above 30 ppm, odor described as sweet or sickeningly sweet
2–5	Prolonged exposure may cause nausea, tearing of the eyes, headaches, or loss of sleep. Airway problems (bronchial constriction) in some asthma patients
20	Possible fatigue, loss of appetite, headache, irritability, poor memory, and dizziness
50–100	Slight conjunctivitis ("gas eye") and respiratory tract irritation after 1 hour. May cause digestive upset and loss of appetite
100	Coughing, eye irritation, loss of smell after 2–15 minutes (olfactory fatigue). Altered breathing and drowsiness after 15–30 minutes. Throat irritation after 1 hour. Gradual increase in severity of symptoms over several hours. Death may occur after 48 hours
100–150	Loss of smell (olfactory fatigue or paralysis)
200–300	Marked conjunctivitis and respiratory tract irritation after 1 hour. Pulmonary edema may occur from prolonged exposure
500–700	Staggering, collapse in 5 minutes. Serious damage to the eyes in 30 minutes. Death after 30–60 minutes
700–1000	Rapid unconsciousness, "knockdown" or immediate collapse within 1–2 breaths, breathing stops, death within minutes
1000–2000	Nearly instant death

Source: US (OSHA)/Occupational Safety and Health Administration (OSHA)/Public Domain.

death after one or two breaths. Deaths have been recorded to have occurred at concentrations around 500 ppm, and nearly instant death occurs at concentrations of 1,000–2,000 ppm.

Immediate loss of consciousness has occurred at fairly low levels, as low as 250–300 ppm. Once the victim becomes unconscious and is immobilized, they are unable to leave the area or escape from the toxic gas. Rescues have occurred, and when the victim is moved to fresh air and resuscitated, they have been known to survive. However, deaths occurred when the victim was working alone, was overcome, and died due to H_2S asphyxiation. When working in a high-H_2S area, we should not be working alone; there should be someone readily available who can call for help if someone is exposed and is overcome and collapses.

I'll talk about this later, but another hazard is that we all have a human instinct that when we see a workmate or associate collapse, we want to help and rescue the individual. When this happens in an H_2S environment, the would-be rescuer may also be killed by rushing into the high-H_2S environment without proper training and adequate rescue equipment, including self-contained breathing apparatus. This results in not one casualty, but frequently others succumb to the presence of hydrogen sulfide.

Toxicity is the first reason that there are many deaths attributable to poisoning by hydrogen sulfide. The point is that this is a devious gas; it deadens the sense of smell, and the victim is no longer aware that they are in a dangerous environment. At higher concentrations, the victim can be killed with only one or two breaths. When the victim is exposed to concentrations above about 100–150 ppm, the olfactory nerves are deadened. Then, the only way a victim will know this deadly gas is present is with a personal hydrogen sulfide gas detector or a fixed hydrogen sulfide detector with audible and visual alarms. This is extremely important, and all personnel working in or around areas with high concentrations of H_2S should always wear an H_2S detector. This means a detector should be worn during every visit to the field or onto the unit where the H_2S is present in high concentrations. Appendix 2 provides many examples of workers who were not wearing an H_2S detector and have been killed, and unfortunately, these incidents are still happening today.

The Material Safety Data Sheet (SDS) provides more details on the toxicity of H_2S. Please see the Additional Resources section for a link to an example of the Hydrogen Sulfide SDS.

Hydrogen sulfide toxicity is the primary reason hydrogen sulfide is so deadly.

The Impact on the Olfactory Nervous System

The odor receptors, present in extremely large numbers (millions), are located on olfactory receptor cells in the back of the nasal cavity and line the roof and sides of the nasal cavity. They form the olfactory epithelium or the layers of cells that line the hollow nasal organ. The olfactory receptor cells are made up of three types of cells. Olfactory receptor cells are composed of tiny hair-like structures called cilia that detect odor molecules, and there are the supporting cells and stem cells. The olfactory nerve transmits odor information to the brain. The location of the olfactory receptors is in an area where the receptor cells are vulnerable to damage and constantly at risk of exposure to air and other airborne irritants.

One of the most damaging chemicals is the risk of exposure to hydrogen sulfide. At low concentrations, the sense of smell makes hydrogen sulfide easily detectable. It smells like rotten eggs. However, as the concentration increases to 100–150 ppm, the hydrogen sulfide paralyzes the olfactory receptors and kills the sense of smell. Of course, hydrogen sulfide is a dangerous gas because it leaves the victim with the sense that the deadly gas has dissipated and is no longer a threat. When exposed to H_2S, even at low concentrations for an extended period or periodically over time, the sense of smell becomes desensitized – a condition referred to as olfactory fatigue. Continued exposures can also result in more serious human health impacts, such as conjunctivitis, pulmonary edema, and even death. The point is that when H_2S becomes highly toxic, our senses, or our warning signs, are deadened at about this concentration.

Medical research has also indicated that hydrogen sulfide exposure may impact the function of the carotid body in regulating the levels of oxygen, carbon dioxide, and pH in the blood. This organ also functions critically to regulate blood pressure and heart rate. The carotid body is a small cluster of cells located at the division of the carotid artery in the neck and acts as a chemoreceptor to monitor and regulate the levels of these bodily functions. Research indicates that exposure to H_2S may interfere with these life-supporting functions.

The second reason hydrogen sulfide is so deadly is its impact on the olfactory nervous system and its ability to deaden the sense of smell.

Hydrogen Sulfide Absorption in Water

Another factor with hydrogen sulfide is that it is readily absorbed into water, such as wastewater. Hydrogen sulfide is about 10 times more absorbable into water than carbon dioxide. So, how does this contribute to its toxicity? I am sure that most of the readers of this book have opened a carbonated soft drink can or bottle after it was shaken. If so, you were likely sprayed by the rapid release of the carbon dioxide, which carried the soft drink liquid and sprayed it onto your face or clothing. This same thing can happen when water has absorbed hydrogen sulfide, except that it is the H_2S that is sprayed onto the victim. Water can rapidly absorb a tremendous amount of hydrogen sulfide. When it is disturbed, or when the water containing H_2S comes into contact with an acid, the H_2S will be rapidly, possibly violently, released in a cloud of very toxic hydrogen sulfide gas. The H_2S will be released quickly and can overwhelm the victim(s). Consider what can happen when an unsuspecting municipal system worker enters a sewer manhole to replace a pump and disturbs the wastewater.

Heavier Than Air

Hydrogen sulfide, by its nature, is a gas that is heavier than air. This means that when the release of H_2S occurs, the gas will quickly settle to the ground or into underground excavations or lower elevations, including confined spaces with little or no air circulation. The gas will concentrate in these areas, resulting in higher concentrations of H_2S, and can become a trap for people working in or near these environments. This has resulted in many deaths where the victims were not aware the gas had accumulated and concentrated in the lower elevations. This makes H_2S very dangerous, for example, for municipal workers who are performing maintenance on sewer systems or waste treatment plants in below-grade manholes or excavations or for oil and gas workers performing maintenance of pipelines, pumps, and other equipment also located in pump pits or other locations below grade.

When H_2S concentrates in enclosed spaces at or below grade, the concentration is significantly increased, increasing the deadly nature of the gas. The relative density of hydrogen sulfide relative to air is the third reason that there are many deaths attributable to poisoning by hydrogen sulfide.

The density of hydrogen sulfide and its ability to settle into low places and increase its concentration is the third reason H_2S is so deadly.

Flammability

One definition of flammability is the flash point. Hydrogen sulfide is a flammable gas at room temperature and has a flash point well below atmospheric temperature. The flash point of hydrogen sulfide is −116 °F (−82 °C). Therefore, it is a very flammable gas easily ignited by arcs or sparks, such as open flames, hot work, burning or welding, or arcing electrical equipment.

It is not only flammable but also has a wide flammable range, sometimes also referred to as the explosive range. The flammable range is defined as the ratio of fuel to air that will support combustion. If the percentage of the fuel is too low to support combustion, the mixture is said to be too lean. If the percentage of the fuel is too high, it is too rich to burn. The flammable range for hydrogen sulfide is 4.3% LEL (lower explosive limit) to 45.5% UEL (upper explosive limit), which is a very wide explosive or flammable range and is, therefore, much easier to ignite. This means that when a release of H_2S does occur, it is more than likely to be within the flammable range, and an explosion or fire can occur if an ignition source is nearby. The flammable range can be found on the Material SDS.

For comparison purposes, the flammable range for propane, which we all know is a very flammable and explosive gas, is 2.1% LEL to 9.5% UEL, which is typical for most hydrocarbons. When compared to the flammable range for hydrogen sulfide, it is easy to recognize the wide range and the even more hazardous nature of H_2S.

When H_2S is involved in a fire, another deadly gas is generated. When hydrogen sulfide (H_2S) burns, it emits sulfur dioxide (SO_2) gas, which has some of the same exposure characteristics as H_2S gas.

The flammability of hydrogen sulfide is the fourth reason that so many deaths have occurred with H_2S.

Chapter 3: Key Lessons Learned

Hydrogen sulfide (H_2S) is a colorless gas with a very distinct rotten egg odor; in low concentrations, it is very noticeable due to this objectionable odor. However, it is a "devious" gas in that at concentrations of between 100 and 150 ppm, it paralyzes the olfactory nerves, and the victim can no longer smell it. The victim will not notice that this has occurred and may believe the gas has disappeared or dissipated and is no longer a hazard.

This characteristic makes H_2S a very dangerous gas and is one reason why there are so many deaths involving exposure to H_2S. The loss of smell occurs at 100–150 ppm; at 500–700 ppm, the victim may be staggering and collapsing after about 5 minutes. Exposures of 700 ppm can result in unconsciousness and death after one or two breaths. Deaths have been recorded to have occurred at concentrations around 500 ppm, and nearly instantaneous death occurs at concentrations of 1,000–2,000 ppm. The Immediately Dangerous to Life and Health (IDLH) level is 100 ppm.

At concentrations above 100–150 ppm, the only way a victim will know this deadly gas is present is with a personal hydrogen sulfide gas monitor or a fixed hydrogen sulfide monitor with audible and visual alarms. All personnel working in or around areas where they could potentially be exposed to hydrogen sulfide should always wear an H_2S detector and immediately leave the area when the detector alarms.

Hydrogen sulfide is heavier than air and will settle into areas below grade, such as excavations, pits, pump pits, or pits for other processing equipment, including piping or manifolds. It will concentrate into deadly concentrations in these locations with little natural ventilation.

Hydrogen sulfide is a toxic gas that will kill at relatively low concentrations and is flammable and explosive. It has a very low flash point (well below atmospheric temperature) and a wide flammable range, which means that if a release occurs, it will most likely be in the flammable range and is easily ignited by a nearby ignition source such as welding, a passing vehicle, or static electricity.

Chapter 3: Incident(s) Related to the Key Lessons Learned

Incident due to the Flammability of Hydrogen Sulfide

Information in this report is credited to the US Occupational Safety and Health Administration (OSHA) and is taken from the OSHA report database.

A fire occurred on May 14, 2022, when a welding torch was being used to seal leaks on a storage tank. The Midland County Fire Marshal reported that the truck explosion was approximately 2.5 miles north of SH-158 on FM 1788 around 9:45 a.m. this morning (May 14, 2022). Three workers were burned in the accident. One person was transported to Midland Memorial Hospital and two people were transported to a hospital in Lubbock by a helicopter. Midland and Odessa authorities responded to the scene.

All three employees were from LCH Trucking LLC and were working near a trailer tank using a welding torch to seal leaks on the tank when they accidentally ignited hydrogen sulfide (H_2S) from the leaking tank, creating a fireball. All three employees were hospitalized to treat second- and third-degree burns. Unfortunately, one of the employees died from his injuries on May 23, 2022.

The company received the following citations from the US OSHA:

> 29 CFR 1910.252(a)(2)(xiii)(A): The employer did not recognize its responsibility for the safe usage of cutting or welding equipment, based on fire potentials of plant facilities, and establish areas for cutting and welding, and procedures for cutting and welding, in other areas: On or about May 14th, 2022, employees engaged in the use of oxygen and acetylene cutting torch operations were exposed to explosion hazards as the employer failed to establish areas for cutting and welding procedures.
>
> 29 CFR 1910.252(a)(3)(i): Welding or cutting was performed on used drums, barrels, tanks or other containers before they were cleaned so thoroughly as to make absolutely certain there were no flammable materials present or any substances which when subjected to heat or any substances such as greases, tars, acids or other material which when subjected to heat might produce flammable or toxic vapors. On or about May 14th, 2022, employees engaged in using sources of ignition to remove bolts from a nipple valve on tanker trailer were exposed to explosion hazards.

Additional Resources

US Occupational Safety and Health Administration, Effects of Hydrogen Sulfide H_2S, https://www.osha.gov/sites/default/files/2018-12/fy15_sh-27664-sh5_Confined_Space_Handout_Effects_of_ H2S.pdf

US Occupational Safety and Health Administration, Hydrogen Sulfide, https://www.osha.gov/hydrogen-sulfide/hazards

Agency for Toxic Substances and Disease Registry, Division of Toxicology and Human Health Sciences, Hydrogen Sulfide – ToxFAQs™, CAS # 7783-06-4, https://www.atsdr.cdc.gov/toxfaqs/tfacts114.pdf

Air Gas, An Air Liquid Company, Safety Data Sheet, Hydrogen Sulfide, https://www.airgas.com/msds/001029.pdf

New Jersey Department of Health, Hazardous Substance Fact Sheet – Right to Know, https://nj.gov/health/eoh/rtkweb/documents/fs/1017.pdf

US Environmental Protection Agency Washington, DC, https://iris.epa.gov/static/pdfs/0061tr.pdf

TOXICOLOGICAL REVIEW OF HYDROGEN SULFIDE (CAS No. 7783-06-4), Concise International Chemical Assessment Document 53, World Health Organization Geneva, 2003, HYDROGEN SULFIDE: HUMAN HEALTH ASPECTS, https://iris.who.int/bitstream/handle/10665/42638/9241530537.pdf

Occupational Safety and Health Administration (OSHA), Inspection: 1598093.015 – LCH Trucking, LLC, https://www.osha.gov/ords/imis/establishment.inspection_detail?id=1598093.015

US Occupational Safety and Health Administration (OSHA), OSHA Fact Sheet – Procedures for Atmospheric Testing in Confined Spaces1, https://www.osha.gov/sites/default/files/publications/atmospheric_test_confined.pdf

US Occupational Safety and Health Administration (OSHA), OSHA Occupational Chemical Database HYDROGEN SULFIDE, (H_2S Properties and Characteristics), https://www.osha.gov/chemicaldata/652

National Library of Medicine, National Center for Biotechnology Information, Anatomy, Head and Neck: Carotid Bodies, https://www.ncbi.nlm.nih.gov/books/NBK562237/

National Library of Medicine, National Center for Biotechnology Information, Hydrogen sulfide activates the carotid body chemoreceptors in cat, rabbit and rat ex vivo preparations, https://pubmed.ncbi.nlm.nih.gov/25583661/

National Library of Medicine, National Center for Biotechnology Information, Long-term effects on the olfactory system of exposure to hydrogen sulphide, https://pubmed.ncbi.nlm.nih.gov/10450248/

Amy Forsgren, Markus Holmberg and Dr. Per Hedmark, Xylem Inc., How to protect against hydrogen sulfide when entering a sewer, https://www.xylem.com/en-us/making-waves/water-utilities-news/how-to-protect-against-hydrogen-sulfide-when-entering-a-sewer/#:~:text=Why%20does%20H2S%20form%20in,velocity)%20of%20sewage%20in%20pipes

National Library of Medicine, The National Center for Biotechnology Information, Hydrogen Sulfide Toxicity, https://www.ncbi.nlm.nih.gov/books/NBK559264/

Chapter 3: End-of-Chapter Quiz

1. Hydrogen sulfide (H_2S) is typically a ________ and ________ toxic gas that in low concentrations has a very distinct ________ ________ odor and in low concentrations it is very noticeable due to this objectionable ________.

2. What is significant about hydrogen sulfide concentration of 0.01–1.5 ppm? Please select the most correct answer from the following:
   a. This is the concentration when the odor is noticeable to most people.
   b. There is really nothing significant about this concentration of H_2S.
   c. 0.01–1.5 ppm is the concentration when the olfactory nerves are paralyzed, and the odor is no longer noticeable.
   d. At this concentration, it is time to evacuate all personnel from the immediate area.

3. At concentrations between ________ and ________ ppm, it paralyzes the ________ nerves, and the victim can no longer smell it.

4. Hydrogen sulfide is ________ than air and will settle in ________ places.

5. Concentrations between ________ and ________ ppm result in almost instantaneous death.

6. Hydrogen sulfide is toxic, but it is also flammable, and at certain concentrations it can be ________.

7. Hydrogen sulfide is not only ________, but it has a wide ________ ________, meaning that it is much easier to ignite.

8. Hydrogen sulfide is easily ignited by a nearby ignition source such as ________, a passing ________, or static ________.

9. When present at concentrations above about 100–150 ppm, the only way a victim will know that this deadly gas is present is with a personal hydrogen sulfide gas ________ or a fixed ________ ________ monitor with an alarm.

10. We all have an ingrained instinct that when we see a workmate or associate collapse, we have the human urge to ____________.

11. The flammable range for hydrogen sulfide is ________% LEL to ________% UEL.
Please select the correct ratio of fuel to air for hydrogen sulfide.
a. 6.2% LEL to 21% UEL
b. 4.3% LEL to 45.5% UEL
c. 2.5% LEL to 51.7% UEL
d. 4.4% LEL to 11.5% UEL

12. Where can we find the documented characteristics of hydrogen sulfide?

13. Hydrogen sulfide has a very __________ flash point and is easily ignited.

4

The Hazards of Hydrogen Sulfide

Believe it or not, we still have more than one system of hazard identification for chemicals in the United States and worldwide. The key systems in the United States and some other countries are the CAS (Chemicals Abstracts Service) and the UN (United Nations) numbering system. In the United States, the UN is also called DOT (Department of Transportation) on the Safety Data Sheets (SDS).

In the United States, the UN and DOT international hazard code for hydrogen sulfide is UN 1053. The CAS number is 7783-06-4. If you type either of these codes into almost any computer search bar, the personnel safety precautions for hydrogen sulfide, including access to the Hydrogen Sulfide SDS, should immediately appear. If it doesn't, you can visit the EPA Cameo Chemicals database at https://cameochemicals.noaa.gov/. When the Cameo Chemicals database opens, simply search for the chemical by name (hydrogen sulfide), CAS number (7783-06-04), or UN/NA number (UN 1053).

The SDSs provide you with information on the material, including other names and identification numbers for the same chemical. It also describes some of its uses, the hazards of exposure or contact with the chemical, and precautions for working with or around it. The SDS covers information on first aid and firefighting measures, and other information on actions in the event of an accidental release of the chemical. The SDS also covers safe handling and storage procedures, and exposure prevention and control procedures. The SDS goes further and provides detailed information on the physical and chemical properties of the chemical, its stability and/or reactivity. It also provides information on the chemical's toxicology, ecological effects, and disposal and transport considerations.

Hydrogen sulfide becomes a serious risk to workers when it occurs or is concentrated in confined spaces or areas with little or no air circulation. Since it occurs naturally in sewers or manure pits, around or in oil and gas wells, and in natural formations in the ground that may be confined spaces, these locations are

Hydrogen Sulfide (H_2S) Hazards: Risk Recognition and Prevention, First Edition. M. Darryl Yoes.

particularly hazardous for workers. Many sanitation workers have been exposed to and have died while cleaning or working in or near sewers, septic tanks, or municipal wastewater treatment plants due to the unexpected presence of hydrogen sulfide. A listing of these incidents that occurred in the United States is provided in Appendix 2, "OSHA Hydrogen Sulfide Injury & Fatality Data."

According to the Bureau of Labor Statistics (BLS), hydrogen sulfide is a primary cause of asphyxiation in the United States. They reported that H_2S deaths accounted for 46 fatalities in the workplace between 2011 and 2017. This reinforces the statement made in the Note to Readers that these deaths are still happening today. More recently, the Occupational Safety and Health Administration (OSHA) reported there were 60 fatalities related to hydrogen sulfide poisoning in the oil and gas industry in the United States between 2012 and 2022.

Poisoning due to H_2S has been ranked as a leading cause of death in the oil and gas industry, as well as in other industries in the United States and worldwide. In the United States, this follows only transportation and contact with objects and equipment. This is happening abroad as well, although data is lacking on the frequency of occurrence.

To illustrate that this trend is continuing, there were 29 reported deaths due to exposure to hydrogen sulfide during the period 2019–2023 in the United States alone. During this same period, there were 21 injuries reported, also as a direct result of H_2S poisoning. Most of these injuries required hospitalization. In 2022, a particularly bad year, a gas plant operator and his wife died in a tragic event in the Permian Basin near Odessa, Texas. This incident is discussed in more detail in Chapter 7, "What Has Happened/What Can Happen Again." Another tragic H_2S incident occurred in September 2022, when an oilfield worker fatally inhaled H_2S gas near a sump pit at a waste treatment and disposal facility, also in the Permian Basin, operated by Production Waste Solutions near Orla, Texas.

More recently, on October 10, 2024, local media reported a hydrogen sulfide release at a Pemex facility in the Houston suburb of Deer Park, Texas. The local sheriff said the release occurred when workers were working on a flange. The release resulted in the deaths of 2 workers, and 35 others were either treated at the site or were transported to local hospitals for treatment. The release resulted in a shelter-in-place for Deer Park and parts of nearby Pasadena, Texas, where residents were asked to remain indoors and close doors and windows. Again, this incident exemplifies that deaths are still happening with hydrogen sulfide, and we must be more aware and take necessary precautions to prevent additional loss of life. This is exactly what this textbook is all about.

Appendix 2 contains a review of OSHA-reported injuries over the past 30 years, including 179 deaths and 124 injuries, including 101 hospitalizations. This includes the following H_2S incidents (some of these are covered in more detail in Chapter 7 "What Has Happened/What Can Happen Again?").

- The deaths of the Alcorn Water Flood Station operator and his wife, both due to hydrogen sulfide poisoning.
- The deaths of two brothers and two employees who were replacing a sump pump in an underground lift station, which collected drainage and leached water from a construction debris landfill. One brother entered the excavation and immediately collapsed. Others followed him in quick succession as they were attempting to rescue him. A total of four deaths occurred at this site due to the deadly hydrogen sulfide gas.
- The transfer of sodium hydrogen sulfide into a tank containing dilute sulfuric acid results in a major release of hydrogen sulfide and one fatality.

If you are interested in seeing the OSHA database for hydrogen sulfide injuries and fatalities, it is available as Appendix 2, as it existed when this book was written. You can also search the OSHA database for updated incident data starting with the following link and updating the search with your search criteria, such as SIC code, incident date range, etc.

https://www.osha.gov/ords/imis/AccidentSearch.search?acc_keyword=%22Hydrogen%20Sulfide%22&keyword_list=on

The following lists several other H_2S incidents not reported in the OSHA database. This indicates that even though the OHSA data is very comprehensive and good for our review purposes, it is more than likely not a true representation of the total number of H_2S incidents and the number of lives lost due to exposure to hydrogen sulfide.

- The Miami Herald reported the deaths of three brothers in a manure pit on an Ohio farm on August 10, 2021, due to the accumulation of toxic gases, including H_2S. This was not found in the OSHA data (3 fatalities). Another great report in the Michigan Farm News, titled "Recent Ohio Fatalities: A Tragic Reminder of Manure Storage Dangers," not only covers this incident but also highlights the hazards of working around manure pits. See the links in the Additional Resources section for these two reports on this accident and the valuable lessons learned.
- On October 12, 2011, two brothers died due to hydrogen sulfide exposure while cleaning a drainage from a waste landfill. One brother fell into an excavation and collapsed, and the other attempted to rescue him. Both died due to H_2S exposure. See the Additional Resources section for a report on this accident. I later realized that this incident was included in the OSHA database, although OSHA reported one fatality and did not report that the victims were brothers.
- On September 2, 2005, the Acute Communicable Disease Control (ACDC) unit was notified of three deaths occurring in the propeller room of a cruise ship docked in the Port of Los Angeles. Initial reports suggested that the

deaths were related to hydrogen sulfide poisoning while crewmembers attempted to fix a pipe in the propeller room on the ship. Hazmat obtained an environmental air sample at 11:00 a.m. on September 3, 2005, five hours after the incident in the propeller room, measuring 100 ppm of H_2S (this incident resulted in 3 fatalities and as many as 19 injuries). This accident was not found in the OSHA database, possibly because it was a marine incident.

Why Is Hydrogen Sulfide so Deadly?

Hydrogen sulfide (H_2S) is an extremely dangerous and elusive gas!

It is highly toxic, flammable, and explosive. Exposure to high concentrations, above about 1,000–2,000 ppm, can result in instantaneous unconsciousness and death to victims. Continuous exposure to concentrations as low as 500 ppm over longer periods can also result in death. Exposure to 100–150 ppm can paralyze the olfactory nerves, which deadens the sense of smell. This leaves the victim thinking that the gas has dissipated and that they are no longer exposed. Hydrogen sulfide inhibits the cytochrome oxidase enzyme system, which is required to maintain cellular respiration and utilize available oxygen.

Hydrogen sulfide is also known for its "knockdown" characteristics. This happens when the victim inhales a breath of H_2S at about 700–1000 ppm. The effect is almost instantaneous unconsciousness. The victim quickly collapses and falls, frequently to their death. For example, a truck driver opens the hatch on his tank truck, inhales the vapors containing H_2S, and then collapses and falls from the truck, frequently sustaining a serious injury or even death.

H_2S is rapidly absorbed by the lungs and transferred to other human organs, resulting in a swift impact on the body. Exposure to low concentrations over an extended period can also lead to long-term health effects. For example, exposure to lower concentrations can result in neurological effects, affects respiration, and is an eye irritant.

- It is a flammable vapor and explosive and has a wide flammable range, making it easily ignited. Concentrations of H_2S have exploded due to the very low flash point and wide flammable range (ratio of H_2S and air).
- It is toxic, even at low concentrations.
- It deadens the sense of smell at around 100–150 ppm, and the victim may believe it is no longer a threat.
- At levels of only 500–700 ppm, it can result in loss of mobility and death.
- It is heavier than air and will travel along the ground and can accumulate and concentrate in low areas such as pits, sewers, underground vaults, etc.
- It will quickly and completely disable both victims and rescuers. Many rescuers have been killed while attempting to rescue a fallen workmate. We will discuss this in more detail in later chapters.

- It can kill instantly at concentrations of around 1,000 ppm.
- It is also found in swamps and bogs due to decaying vegetation, and contact and exposure may be completely unexpected.

Hydrogen Sulfide in Oil and Gas Operations

According to the BLS, hydrogen sulfide is one of the leading causes of workplace gas inhalation deaths in the United States. Between 2011 and 2017, 46 deaths were due to H_2S. Unfortunately, this is especially true for oil and gas operations, as evidenced by the Pemex Refinery incident in October 2024.

Oil and natural gas drilling and petroleum refining workers may be exposed because hydrogen sulfide may be present in oil and gas deposits. In refining, it is also a by-product of the desulfurization process of these fuels. Working in these areas and conditions increases a worker's risk of overexposure to hydrogen sulfide. Workers must receive specialized training and personal protective equipment (PPE) to protect themselves against these hazards. There will be much more to come on this in subsequent chapters.

Petroleum refineries typically have foul amine, amine containing high concentrations of hydrogen sulfide, and hydrogen sulfide circulating in piping systems at concentrations in the 80–90% range or 800,000–900,000 ppm. Of course, one breath of H_2S at this concentration results in instantaneous knockdown and death.

Confined spaces or areas with limited or no air circulation can accumulate hydrogen sulfide, and the vapors can become concentrated and a serious health hazard for anyone who enters the confined space. This can happen in excavations or pits, especially manure pits, manholes, tunnels, and any enclosure where the hydrogen sulfide can build up to dangerous levels. For example, farm and ranch workers have been exposed and killed while cleaning manure storage tanks or working in manure pits. Another hazard is anywhere that has marshy landscapes or where vegetation or other organic matter can be broken down by bacteria, creating hydrogen sulfide.

The common mode of exposure for humans is inhalation of H_2S vapors. The lungs rapidly absorb the H_2S gas. It is easily detected in low concentrations by its distinctive rotten egg odor; however, higher concentrations paralyze the olfactory nerves, defeating the sense of smell. Victims may assume that since they can no longer smell the H_2S, it has dissipated or gone away. Unfortunately, the gas may be even stronger and result in the death of the victim or victims.

Refinery or oil field operations personnel are particularly vulnerable to the hazards of hydrogen sulfide as they are working around it almost continuously. They are the ones who are responding to alarms, pump trips, and other similar issues on an ongoing basis. If they are working at a facility that does not provide toxic gas detectors, or if management fails to enforce the use of these protective devices, the workers may not have any warning that they are being exposed to H_2S or are in danger of being killed. More information is forthcoming on the effectiveness of these devices, but they are not effective if they are not being used in the field; for example, if they are left in the control room, office, or on the seat of their vehicle. Detectors should be worn on the person, near the breathing zone, at all times in the field or on the process unit.

Flammability of Hydrogen Sulfide

As outlined earlier, hydrogen sulfide is a very toxic gas. However, this is not the only hazard; it is also a highly flammable gas and has a wide flammable range. The flammable range, sometimes called the explosive range, is 4.3–45.5% in air, which is very wide. When compared to propane, a very flammable and volatile petroleum vapor, its flammable range is only about 2% LEL to about 9.5% UEL. Hydrogen sulfide is also explosive and has resulted in fires, explosions, and deaths, especially in the petroleum refining industry and in the gas and oil fields, where high concentrations of H_2S are present during routine operations. This means that unexpected releases of H_2S must always be a high priority, especially during maintenance operations or when equipment that normally contains H_2S is being opened for inspection or maintenance. More on this to come in the chapters "What Has Happened/What Can Happen Again" and "How the Hazards of Hydrogen Sulfide Are Controlled."

Hydrogen Sulfide Is Heavier Than Air

Hydrogen sulfide (H_2S) is significantly heavier than air and will quickly settle into low places. Due to this characteristic, the gas will concentrate when it settles, increasing its toxicity. H_2S has a specific gravity of 1.19, meaning it is 1.19 times denser than air. The molecular weight of H_2S is 34.08 g/mol compared to 29 g/mol for air. This makes H_2S about 19% denser than air.

This means that when H_2S is released into the air, it will quickly settle into areas at lower elevations, such as pump pits, open sewers, open wastewater drains, open excavations, and so on. Also, as the gas concentrates, it becomes increasingly toxic and much more dangerous if inhaled.

Hydrogen Sulfide Can Readily Be Absorbed into Water

Hydrogen sulfide is readily absorbed into water at concentrations about ten times greater than carbon dioxide. This means that standing or stagnant water can contain a massive concentration of hydrogen sulfide, which has been absorbed into the water and is just waiting for the water to be disturbed or stirred. When this happens, the hydrogen sulfide is suddenly released into the surrounding areas where unsuspecting personnel may be working, unaware of its presence. Massive amounts of hydrogen sulfide can also be instantly released if stagnant water comes into contact with any acid or low-pH water, creating a large H_2S vapor cloud. When H_2S is absorbed into water, it can also form sulfuric acid (H_2SO_4), which can also be corrosive to metal and concrete piping and structures.

Personnel Symptoms of Exposure to Hydrogen Sulfide

Exposure to hydrogen sulfide can occur quickly and without warning. The victim may smell the odor of rotten eggs; however, this goes away quickly with higher concentrations of H_2S. The symptoms may occur suddenly and without warning.

Identifying these symptoms quickly and ensuring prompt medical attention is critical to prevent long-term health effects. The symptoms can be very mild but quickly become severe or acute. Table 3.1 in Chapter 3 illustrates how these symptoms relate to the amount of exposure to toxic hydrogen sulfide. The toxic gas, when first noticeable, smells like rotten eggs. This happens at extremely low concentrations, from 0.01 to about 1.5 ppm. However, at higher concentrations, it becomes extremely dangerous and can kill. This is because it deadens the olfactory nerves at about 100–150 ppm; the victim no longer smells it and is led to believe that the gas has dissipated.

A Hydrogen Sulfide Case Study

Let's describe what happens to a victim when exposed to hydrogen sulfide. Even though these symptoms are spelled out in Table 3.1, I believe the symptoms are more real and tangible when we talk about what is happening to the victim. The story below is a real event. However, I have made editorial changes, and you will quickly recognize that I have fabricated parts of this, and the names have been changed. The principal change is that I have you participating in this incident. However, this is a real event, and I wanted this discussion to be realistic.

> You and your best friend, Jim, were fortunate to land a job working for an oil and gas company in the gas fields near Watford, North Dakota. You and Jim have worked together for about 12 years, and you grew up together in the oil fields. Not only do you work together, but you have also become best friends away from the job. You and your family had a barbecue with Jim and his family last week, and you guys do this frequently. You and

Jim attended awareness training on hydrogen sulfide a couple of years ago, where you learned about the symptoms of H_2S and how critical it is to know these symptoms and to prevent exposure to hydrogen sulfide to prevent injury, long-term health effects, or worse.

It was a cold evening in February, and Jim was working on a shift with another teammate and good friend, Bob. Jim was working alone in an oil and gas pumping operation and was in a small well pump shack, about 9 × 10 feet (a small building, about 90 square feet), attempting to restore the pumping operations. Due to the cold weather, the door to the shack was closed, meaning there was very little air circulation in the pump shack. Jim connected a hose from the pump discharge to the flare. He then ran the opposite end of the hose into an empty metal bucket inside the pump shack, attempting to get circulation started on the pump. When he opened the valve to the hose, vapor was released with a small amount of liquid. Jim quickly smelled the H_2S. It is very noticeable at about 0.01–1.5 ppm and smells like rotten eggs. The odor becomes very offensive at about 30 ppm.

Jim continued working, and when the concentration of H_2S reached about 100–150 ppm, the smell quickly disappeared. The H_2S had deadened Jim's olfactory nerves, and although it was still present, Jim could no longer smell it. This is why I refer to H_2S as a devious or deceitful gas; it tricks the victim into believing it has dissipated and is safe to continue working.

As Jim continued to focus on the task at hand, to get the pump to start pumping and restore pumping from the tank, the concentration of H_2S continued to increase. At concentrations of 200–300 ppm, Jim would have severe respiratory tract irritation and eye irritation (like conjunctivitis or pink eye). Remember, at these concentrations, Jim no longer has any other indications that the deadly gas is present and does not recognize that the concentration is increasing.

Suppose the exposure at 200–300 ppm lasts for an hour. In that case, he will experience more severe conjunctivitis and respiratory tract irritation, and he may experience pulmonary edema, and fluids would accumulate in the lungs. At 500–700 ppm, he will stagger and collapse in about 5 minutes. If exposure continues at this concentration, death will occur in about 30–60 minutes.

Rapid unconsciousness occurs at 700–1,000 ppm, and at this level, he would collapse in 1–2 minutes, and death would occur within minutes. Concentrations at or around 1,000–2,000 ppm result in almost instantaneous death.

Bob had been trying to contact Jim, but Jim wasn't responding to his radio or cellular telephone. Bob became concerned and drove to the pump shack where Jim was last known to be working. He found that the door to the shack was closed, and Jim was inside. Jim was found to have collapsed and unresponsive. Bob also found a hose connected to a pump discharge valve that goes to the flare and an overturned bucket on the floor with a small amount of fluids in the bucket. He noticed condensate and a small amount of produced water at the end of the hose. Jim also had chemical-type burns on several parts of his body.

Bob called for assistance, and Jim was transported to the hospital, where he was pronounced dead due to exposure to hydrogen sulfide.

We don't know the concentration of H_2S that was released in the well pump house or how quickly it was released. However, data from other wells in the area would confirm that the concentration was very high, and if similar concentrations were released here, it would quickly result in Jim's death. Since the report from the officials on this incident failed to mention an H_2S detector or SCBA, we must assume Jim was not wearing either. Jim was found to be deceased due to hydrogen sulfide exposure.

This event was taken from one of the OSHA reports in Appendix 2. It is a real event, although the names have been changed. However, a fatality did occur, very much as described earlier. What did we learn from this discussion?

- Always use the Last-Minute Risk Assessment (LMRA) process to identify job site hazards before starting a new task or changing the process used for executing an ongoing task.
  - Identify the hazards associated with this task.
  - Assess the risks.
  - Define how the hazards or risks can be controlled or mitigated.
  - Act to ensure the mitigation is in place.
  - As the task is implemented, carefully follow the mitigation plan.
  - Evaluate the effectiveness of the mitigation and make changes as necessary.
- Never enter a unit or area designated as a potential H_2S area without an H_2S functioning detector.
  - If it alarms, immediately leave the area. Do not investigate until donning an SCBA.
- Never open equipment in high-concentration H_2S without first donning an SCBA.
- Always be aware when working near H_2S equipment in confined spaces or areas with limited air circulation.
  - In these cases, an SCBA may be required.
- Procedures should be developed for standardized tasks in checklist format, and they should be closely followed.
- If unsure how to safely execute a task, always ask for help!

Chapter 4: Key Lessons Learned

Hydrogen sulfide is a serious risk to workers when it occurs or is concentrated in confined spaces or areas with little or no air circulation. It occurs naturally in sewers or manure pits, around or in oil and gas wells, and in natural formations in the ground, which by their nature may be confined spaces; these locations are particularly hazardous for workers. Many sanitation workers have been exposed to and have died while cleaning or working in or near sewers, septic tanks, or municipal wastewater treatment plants due to the unexpected presence of hydrogen sulfide.

The OSHA reported there were 60 fatalities related to hydrogen sulfide poisoning in the oil and gas industry in the United States between 2012 and 2022, and poisoning due to H_2S has been ranked as a leading cause of death in the oil and gas industry, and in other industries in the United States and worldwide, following only transportation and contact with objects and equipment. Recent incidents at an oil and gas well water pumping station in the Permian Basin near Odessa, Texas, and a refinery in Deer Park, Texas, clearly illustrate that this deadly trend continues today.

Inhalation of hydrogen sulfide vapors is the most common mode of exposure for humans. The lungs rapidly absorb the toxic gas and quickly transfer it to other human organs, such as the brain. It is easily detected in low concentrations by the distinctive odor of rotten eggs. However, it is toxic, even at low concentrations, and it deadens the sense of smell at around 100–150 ppm, and the victim may believe it is no longer a threat. The OSHA 8-hour limit is 20 ppm with a ceiling of 50 ppm with a 10-minute peak exposure. The NIOSH Immediately Dangerous to Life and Health level is 100 ppm. At levels of only 500–700 ppm, it can result in loss of mobility and death. Victim knockdown occurs at about 700–1,000 ppm, resulting in unconsciousness and potential falls. At levels of 1,000–2,000, death is instantaneous.

Hydrogen sulfide is a highly flammable vapor and explosive and has a wide flammable range, making it easily ignited. Concentrations of H_2S can and have exploded when in the flammable range (ratio of H_2S to air). Workers have been killed and severely injured due to a fire involving H_2S.

It is heavier than air and will travel along the ground and concentrate in low areas such as pits, sewers, underground vaults, etc. It will quickly disable both victims and rescuers. Many rescuers have been killed while attempting to rescue a fallen workmate.

Due to decaying vegetation, H_2S is also found in swamps and bogs, and contact and exposure may be completely unexpected. Significant amounts of hydrogen sulfide can be absorbed into stagnant water and unexpectedly released when the water is disturbed or when it comes into contact with any form of acid. H_2S absorption into the water can also form sulfuric acid (H_2SO_4), which can be corrosive to metal and concrete piping and structures.

Preventing unexpected releases of H_2S must always be a high priority, especially during maintenance operations or when equipment that normally contains H_2S is being opened for inspection or maintenance. A facility that handles super-rich H_2S, such as petroleum refinery acid gas streams, pipelines, and systems, demands special precautions anytime inspection or maintenance is carried out on or near these systems. See the chapter "How the Hazards of Hydrogen Sulfide Are Controlled" for additional information.

Anytime we work in or near areas containing H_2S, we should wear H_2S detectors. These can and will save lives if they are continuously checked for functionality and worn properly by operators and others who may be exposed to a release of H_2S. Also, anytime we are working on systems or equipment that contain H_2S at high concentrations, or if the H_2S in the area exceeds the OSHA working requirements (20 ppm), we should wear breathing air protection (SCBA). The use of SCBA can save lives.

The National Library of Medicine and the National Center for Biotechnology Information published an article titled "Hydrogen sulfide exposure in an adult male." In that article, they presented data on the effects of hydrogen sulfide exposure and included how these can be prevented with appropriate PPE. In that article, they reported that 77 of the 80 deaths were thought to be potentially preventable using an H_2S alarm or portable meters.

Chapter 4: Incident(s) Related to the Key Lessons Learned

Incident: Hydrogen Sulfide Being Heaver Than air and Accumulating in High Concentrations in Low Places

US OSHA: "Two employees are killed from overexposure to gas in vault"

On May 25, 2021, two employees from the Northwest New Mexico Regional Solid Waste Authority were conducting maintenance work on a pump above a leachate vault, which is designed to handle the accumulated water and other liquids for the solid waste landfill. A coworker dropped a cellphone inside the vault and left to look for a tool to retrieve the cellphone. Before the coworker returned, the first employee entered the vault to retrieve the phone and was overcome by methane and hydrogen sulfide gas. When the coworker returned, he also entered the vault and was also overcome. Both employees were killed by the hydrogen sulfide gas from the leachate, which is a well-known hazard.

The employees entered the confined space without performing a gas test of the atmosphere, without the required confined space attendant, and without a rescue plan or rescue equipment, basically without any of the critical compliance components of a confined space entry. There were none of the expected precautions such as a toxic gas detector. The OSHA issued 12 citations to the company for this incident.

Additional Resources

US Chemical Safety and Hazard Investigation Board (CSB), Aghorn Operating Waterflood Station Hydrogen Sulfide Release, https://www.csb.gov/aghorn-operating-waterflood-station-hydrogen-sulfide-release-/

California Fatality Assessment and Control Evaluation (CA/FACE) Program, Two Laborers Die from Hydrogen Sulfide Exposure in a Confined Space at an Organic Waste Recycling Facility, https://www.cdc.gov/niosh/face/pdfs/11CA008.pdf

Miami Herald, August 19, 2021: Three brothers die after passing out in manure pit at Ohio farm, authorities say, Read more at: https://www.miamiherald.com/news/nation-world/national/article253418144.html#storylink=cpy, https://www.miamiherald.com/news/nation-world/national/article253418144.html

AGDaily, "Poisonous gases released by stored manure can cause immediate asphyxiation and death," https://www.agdaily.com/livestock/recent-fatalities-in-ohio-a-tragic-reminder-of-manure-storage-dangers/

PennState Extension, Manure Gas Safety, https://extension.psu.edu/manure-gas-safety

Draeger, H_2S (hydrogen sulfide) - Knowledge can save lives, working at the refinery – basic information on the safe handling of hazardous substances H_2S, https://www.draeger.com/Content/Documents/Content/H2S%20Booklet-bk-PDF-9088-en-master-1810-1.pdf

United States Mine Rescue Association, Hydrogen Sulfide Awareness *(A PPT Presentation)*, https://view.officeapps.live.com/op/view.aspx?src=https%3A%2F%2Fminingquiz.com%2Fpowerpoints%2FMine_Gases%2FH2S_Awareness_Rev3.ppt&wdOrigin=BROWSELINK

Wiley Guide to Chemical Incompatibilities, 2nd Edition, Richard P. Pohanish, Stanley A. Greene, https://www.wiley.com/en-us/Wiley+Guide+to+Chemical+Incompatibilities%2C+2nd+Edition-p-9780471721628

The University of Notre Dame, Incompatible Chemicals, https://riskmanagement.nd.edu/assets/444930/incompatible_chemicals.pdf

NBC News, 2 dead after a hydrogen sulfide leak at a Houston-area Pemex facility, https://www.nbcnews.com/news/us-news/2-dead-hydrogen-sulfide-leak-houston-area-pemex-facility-rcna174984

Agency for Toxic Substances and Disease Registry (ATSDR), Medical Management Guidelines for Hydrogen Sulfide, https://wwwn.cdc.gov/TSP/MMG/MMGDetails.aspx?mmgid=385&toxid=67

Canadian Substance Search, https://pollution-waste.canada.ca/substances-search/Substance/SearchByName?Name=hydrogen+sulfide&ExactMatch=false&SearchLang=en

US Occupational Safety and Health Administration (OSHA), Accident Report Detail, Accident Summary Nr: 136332.015, Two employees are killed from overexposure to gas in vault, Accident Summary Nr: 136332.015 -- Report ID: 0653510 -- Event Date: 05/25/2021, https://www.osha.gov/ords/imis/accidentsearch.accident_detail?id=136332.015

Government of Canada, Hydrogen sulfide and your health, https://www.canada.ca/en/health-canada/services/chemicals-product-safety/hydrogen-sulfide.html

The National Library of Medicine, National Center for Biotechnology, Hydrogen sulfide exposure in an adult male, https://pmc.ncbi.nlm.nih.gov/articles/PMC2850187/#CIT26

Michigan Farm News, Recent Ohio fatalities a tragic reminder of manure storage dangers, https://www.michiganfarmnews.com/recent-ohio-fatalities-a-tragic-reminder-of-manure-storage-dangers

US Occupational Safety and Health Administration (OSHA), Oil and Gas Well Drilling and Servicing eTool, https://www.osha.gov/etools/oil-and-gas

Centers for Disease Control (CDC)/National Institute for Occupational Safety and Health (NIOSH), NIOSH Warns: Manure Pits Continue to Claim Lives, https://archive.cdc.gov/#/details?q=manure%20pits%20hydrogen%20sulfide&start=0&rows=10&url=https://www.cdc.gov/niosh/updates/93-114.html00

Centers for Disease Control (CDC)/National Institute for Occupational Safety and Health (NIOSH), Preventing Deaths of Farm Workers in Manure Pits, https://archive.cdc.gov/#/details?q=manure%20pits%20hydrogen%20sulfide&start=0&rows=10&url=https://www.cdc.gov/niosh/updates/upd-07-03-07.html

National Library of Medicine, National Center for Biotechnology Information, Hydrogen Sulfide, https://www.ncbi.nlm.nih.gov/books/NBK219913/

Forensic Science International/Science Direct, Lethal Hydrogen Sulfide poisoning in open space: An atypical case of asphyxiation of two workers, https://www.sciencedirect.com/science/article/abs/pii/S0379073819305341

Fuller DC, Suruda AJ. Occupationally related hydrogen sulfide deaths in the United States from 1984 to 1994. *J Occup Environ Med.* 2000;42:939–42. doi: 10.1097/00043764-200009000-00019. [DOI] [PubMed] [Google Scholar]

Chapter 4: End-of-Chapter Quiz

1. Hydrogen sulfide becomes a serious risk to workers when it occurs or is ____________ in confined spaces or areas with ________ or ____ ______ circulation.

2. The common mode of exposure for humans is by ____________ of the vapors. The H_2S gas is rapidly absorbed by the __________.

3. The victim may assume since they can no longer smell the H_2S that it has __________ or has gone away.

4. Asphyxiation by H_2S can happen in excavations or pits, especially in the following.
   Name as many as possible.

5. Why is hydrogen sulfide so deadly? Please list as many as possible.

6. According to the Bureau of Labor Statistics (BLS), hydrogen sulfide is a primary cause of ____________ in the United States.

7. Petroleum refineries typically have foul amine and hydrogen sulfide circulating in piping systems at concentrations in the 80–90% range or 800,000–900,000 ppm concentration. One breath of H_2S at this concentration results in ______________ death.

8. Hydrogen sulfide occurs naturally in sewers or manure pits, around or in oil and gas wells, and natural formations in the ground which by their nature may be _________ _________, these are particularly ____________ for workers.

9. Sanitation workers have been exposed to and have died while cleaning or working in or near sewers, septic tanks, or municipal wastewater treatment plants due to the unexpected presence of ____________ ____________.

10. What is the flammable range for hydrogen sulfide?

5

The Working Limits for Hydrogen Sulfide

The US Occupational Safety and Health Administration (OSHA) reports that, "Hydrogen sulfide gas causes a wide range of health effects. Workers are primarily exposed to hydrogen sulfide by breathing it (by inhalation of the toxic vapors). The effects depend on how much hydrogen sulfide the victim breathes and for how long." OSHA also states, "Exposure to very high concentrations (of hydrogen sulfide) can quickly lead to death." They also provide information that H_2S is heavier than air, can accumulate in low areas, and that concentrations can increase to very high levels in these low areas. The link to the OSHA Fact Sheet on Hydrogen Sulfide is provided in the Additional Resources section of this chapter.

Exposure standards for the workplace are established by OSHA and NIOSH (National Institute for Occupational Safety and Health), the US agency responsible for studying the effects of hazardous work environments and for making recommendations for the prevention of work-related injuries and illnesses. NIOSH is also the Centers for Disease Control and Prevention (CDC) agency that conducts research, training, and surveillance on worker safety and health.

OSHA and NIOSH have established workplace limits for hydrogen sulfide, including the following. These workplace limits are summarized in Chart 5.1.

- General Industry: Current OSHA PEL: 20 ppm for an 8-hour period, with a ceiling of 50 ppm for a 10-minute MAXIMUM PEAK, with no other exposures during the work period.
- OSHA PEL for Construction and Shipyards is 10 ppm.
- NIOSH-recommended exposure limit (REL) is 10 ppm (15 mg/m^3) with a 10-minute ceiling.
- NIOSH has established an IDLH (immediately dangerous to life or health) of 100 ppm.

Hydrogen Sulfide (H_2S) Hazards: Risk Recognition and Prevention, First Edition. M. Darryl Yoes.

Chart 5.1 Regulatory and recommended short-term exposure limits.

Worker Exposure Limits
NIOSH REL (10-minute ceiling): 10 ppm NIOSH IDLH: 100 ppm OSHA PELs: • General Industry Ceiling Limit: 20 ppm for an 8-hour period. • General Industry Peak Limit: 50 ppm (up to 10 minutes if no other exposure during shift) • Construction 8-hour Limit: 10 ppm • Shipyard 8-hour Limit: 10 ppm

Notes:
REL = Recommended Exposure Limit
PEL = Personnel Exposure Limit
IDLH = Immediately Dangerous to Life and Health
Source: US (OSHA)/Occupational Safety and Health Administration (OSHA)/Public Domain.

It is noteworthy that the Datasheet in the Environmental Protection Agency Cameo Datasheet lists hydrogen sulfide as an AEGL-3 (Acute Guideline Exposure Level 3), which is "Life-threatening health effects or death." The AEGL-3 limit is set for a maximum of 60 minutes at 50 ppm. See the Additional Resources section for a link to this datasheet.

Some industrial facilities adhere to the NIOSH-REL of 10 ppm. Some refining and petrochemical facilities have voluntarily reduced their internal standards to less than 10 ppm.

Chapter 5: Key Lessons Learned

The US OSHA reports that "Hydrogen sulfide gas causes a wide range of health effects that depend on how much hydrogen sulfide the victim breathes and for how long." OSHA also states, "Exposure to very high concentrations of hydrogen sulfide can quickly lead to death." They also warn that hydrogen sulfide will accumulate in high concentrations in low areas.

OSHA and NIOSH have established workplace limits for hydrogen sulfide.

> These workplace limits are OSHA H_2S Personnel Exposure Limit (PEL) of 20 ppm for an 8-hour period, with a ceiling of 50 ppm for a maximum 10-minute peak exposure if there are no other H_2S exposures during the work period. The OSHA H_2S PEL for construction and shipyards is set at 10 ppm.

NIOSH has set a REL of 10 ppm (15 mg/m^3) with a 10-minute ceiling. Most notably, NIOSH has also established an IDLH (immediately dangerous to life or health) of 100 ppm for hydrogen sulfide.

All US sites must comply with OSHA's regulatory requirements for hydrogen sulfide exposures. However, some US industrial facilities adhere to the lower NIOSH-REL of 10 ppm, and some have even voluntarily reduced their internal hydrogen sulfide exposure standards to less than 10 ppm.

OSHA provides data indicating that hydrogen sulfide is heavier than air and will accumulate and concentrate in low areas such as sewers, pump pits, excavations, etc.

Chapter 5: Incident(s) Related to the Key Lessons Learned

Incident Resulting from High-concentration Hydrogen Sulfide Accumulation in Low Areas and Attempted Rescue Without Training and Rescue Equipment (3 Fatalities)

Three employees were working to remove a 24-inch rubber plug from a 2-foot sewer line, and all three were killed by high-concentration hydrogen sulfide gas. At 11:00 a.m. on July 7, 2017, Employee #1 was attempting to dislodge a 24-inch rubber plug from a 2-foot diameter sewer pipe located inside a 24-foot-deep wet well. Workers outside the well were pulling on a 1/4-inch nylon rope that was attached to the 24-inch diameter plug. The plug was lodged inside a T-shaped PVC fitting from the force of the wastewater emptying into the well.

Without conducting any atmospheric testing of the workspace, Employee #1 climbed down the ladder with a crowbar to dislodge the deflated 24-inch diameter rubber plug about 8 feet below the top of the well. He had difficulty releasing the plug with the crowbar and started to make his way up the ladder. He lost consciousness when he was about 2 feet from the top of the well and fell into the 24-foot-deep well. Employee #2 descended the ladder to provide emergency rescue but lost consciousness and went underwater. The wastewater level was about 3 feet deep at this point. Employee #3 climbed down the ladder to provide emergency rescue but lost consciousness. All three workers were asphyxiated by hydrogen sulfide (H_2S) gas.

The OSHA issued 10 serious citations to USA Fanter Corp., Ltd. for violations related to confined space entry, failure to test the atmospheric conditions inside the confined space, failure to protect employees from falls or falling objects, failure to purge the space before entry, failure to provide rescue plans and equipment before entry, and failure to provide relevant safety training.

Additional Resources

US Occupational Safety and Health Administration (OSHA), Safety and Health Topics; Hydrogen Sulfide, https://www.osha.gov/hydrogen-sulfide/hazards

US Occupational Safety and Health Administration (OSHA), OSHA Fact Sheet – Hydrogen Sulfide, https://www.osha.gov/sites/default/files/publications/hydrogen_sulfide_fact.pdf

US Centers for Disease Control and Prevention (CDC), US National Institute for Occupational Safety and Health, https://www.cdc.gov/niosh/index.html

US Centers for Disease Control and Prevention (CDC), US Institute of Occupational Safety and Health (IOSH), US National Institute for Occupational Safety and Health (NIOSH), https://www.cdc.gov/niosh/idlh/7783064.html

Environmental Protection Agency, Cameo Database, Hydrogen Sulfide, https://cameochemicals.noaa.gov/chemical/3625

New Jersey Department of Health, Right to Know – Hazardous Substance Fact Sheet, Hydrogen Sulfide, https://nj.gov/health/eoh/rtkweb/documents/fs/1017.pdf

American Conference of Governmental Industrial Hygienists ("ACGIH®"), https://www.acgih.org/

Abhisam, Free Abhisam Whitepaper, Confined Space Safety Checklist – 5 things to look out for, https://www.abhisam.com/confined-space-safety-5-things-to-look-out-for-checklist/

Chapter 5: End-of-Chapter Quiz

1. Which US Federal agency established the Immediately Dangerous to Life and Health limit for hydrogen sulfide?
   Choose the most appropriate response from the following.
   a. The US Occupational Safety and Health Administration (OSHA)
   b. The US Environmental Protection Agency (EPA)
   c. The US National Institute for Occupational Safety and Health (NIOSH)
   d. The US Centers for Disease Control and Prevention (CDC)
   e. The US National Institute of Health (NIH)

2. The hydrogen sulfide permissible exposure limit is established by the US OSHA. Is it enforceable?
   a. Yes
   b. No

3. Please list the enforceable exposure limits (PEL) established by the US OSHA for hydrogen sulfide in industry.

4. The US Occupational Safety and Health Administration (OSHA) reports that, "Hydrogen sulfide gas causes a wide range of health effects. Workers are primarily exposed to hydrogen sulfide by ____________ it (by ____________ of the toxic vapors)."

5. The effects depend on how ________ hydrogen sulfide the victim breathes and for ______ ________."

6. OSHA also states, "Exposure to very ______ concentrations (of hydrogen sulfide) can quickly lead to ________."

7. What is the Immediately Dangerous to Life and Health limit for hydrogen sulfide?

8. What is the eight-hour working limit for H_2S set by OSHA for:
   a. General industry?
   b. Shipyards?
   c. Construction?

9. The peak H_2S exposure limit set by OSHA for general industry is ______ ppm for a maximum period of ______ minutes, provided there are ______ other exposure during the shift.

10. Some industrial facilities abide by the NIOSH-recommended exposure limit of ______ ppm. Some sites have voluntarily reduced their internal standards to less than ______ ppm.

6

Protecting the Workers and Ourselves Against the Hydrogen Sulfide Hazards

Recognizing the Hazard

When you think about the hazards of hydrogen sulfide and its toxicity, H_2S is primarily an inhalation hazard. It's important to recognize that a high-concentration hydrogen sulfide release, generally concentrations of 1,000 ppm or higher, can kill instantaneously. In this concentration, the victim is knocked down with the first breath, and they must be revived within 2–4 minutes; otherwise, severe brain damage or death will occur. Please note that in a case like this, the victim will not restart breathing on their own; they must be revived. This is why operators, engineers, supervisors, managers, or anyone else should be continuously aware of where they are when out in the field regarding the location of equipment or piping containing hydrogen sulfide in high concentrations. They should be in close contact with someone else who knows their location and activities.

Hydrogen Sulfide Detectors

All personnel, when in the field where H_2S is being processed or is present, even in moderate concentrations (concentrations of 20 ppm or more), should wear an H_2S detector that has been bump-tested against a known standard to ensure it is fully functional. And, anytime the detector goes into alarm status, everyone in the area should immediately evacuate and return to investigate only after properly donning a self-contained breathing apparatus (SCBA). Please remember my analogy in the Note to Readers: use the detector like your seat belt; anytime you go into the field, ensure you have your H_2S detector on your person and near your breathing zone. It may save your life!

Hydrogen Sulfide (H_2S) Hazards: Risk Recognition and Prevention, First Edition. M. Darryl Yoes.

Published 2026 by John Wiley & Sons, Inc.

Many process units, pump houses, and other operator workspaces are designed with hardwired remote H_2S detectors. Likewise, anytime these detectors are in alarm status, this indicates that a release of hydrogen sulfide is occurring. All personnel in the area should evacuate and return to investigate only after wearing a SCBA. Once the release has been stopped and the area has been tested and found clear of H_2S, these restrictions can generally be lifted.

Opening Process Equipment

Anytime personnel are in the process of opening process equipment for insertion of blinds, blind removal, opening access plates, or any other act of OPE (opening process equipment), the vessel or piping system must be verified to ensure that there is no pressure or substantial product present (zero energy). It should also be tested to confirm that it is safe to open. This means the vessel or piping system must be tested at or below 10% of the LEL, 20 ppm or less of H_2S, and 19.5% O_2 or more. Otherwise, all personnel doing the work and anyone else in proximity must be fully outfitted with appropriate personnel protective equipment (PPE), including an SCBA, depending on the gas test results.

Each point of opening for blind insertion, blind removal, access plate removal, or any other vessel or piping system opening must be identified by the process organization. Generally, this is done by hanging a wearproof or weatherproof tag with a unique design. It is only used for this purpose and placed on the flange to be opened. The flanges to be opened should also be marked at each opening point on the P&ID drawing. The process organization is also responsible for ensuring each opening point is at zero energy with no pressure and minimal product presence. To help avoid confusion, each point of opening must be reviewed in detail between the work permit issuer and the work permit receiver or the people doing the work. The craftsmen must understand they are authorized only to open the flanges with this unique tag installed. This review is integral to the unit walk down during the pre-job visit.

Collecting Samples

This will depend on the specific samples being collected and the hazards of those samples. Generally, if the operator is sampling product samples on a hydrotreater, for example, diesel, naphtha, or jet fuels, the samples have been hydrotreated and scrubbed with amines. Therefore, the samples are "sweet" with no hydrogen sulfide present. However, sampling feed to a hydrotreater is totally different. The same is true if the operator is working on a drill rig in West Texas, North Dakota, Alberta, or other areas where crude oil or crude derivatives are known to contain high concentrations of H_2S. In these cases, high concentrations of hydrogen

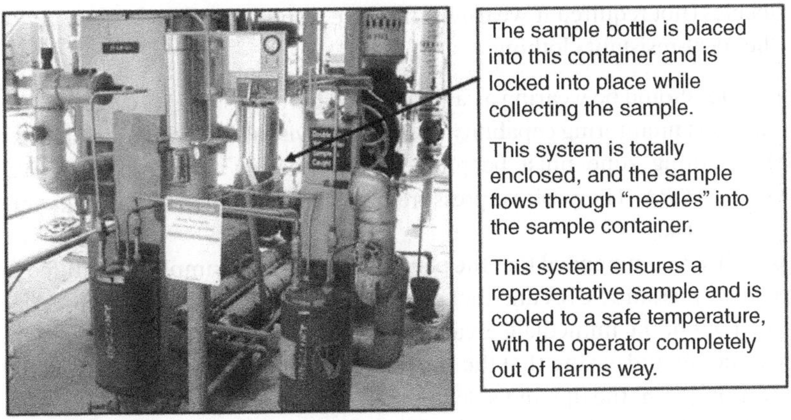

Figure 6.1 Photograph of an engineered closed-loop sample system used for toxic products. Please note that this figure is an older version of a closed-loop sample system. Newer models are readily available from fabricators. Please see the links in the Additional Resources section of this chapter.

sulfide may be present and require either special sampling equipment and/or special safety precautions.

Several manufacturers make available a closed-loop sample system designed to lock a sample container and collect a representative sample while keeping the operator completely out of harm's way. These can be designed for liquids or gases and equipped with internal coolers that cool the sample to a safe temperature before collecting the sample. One of these systems is illustrated in Figure 6.1, and links to several vendors are provided in the Additional Resources section of this chapter.

You will note that this system is equipped with a three-way valve to ensure a continuous flow of the sample, always presenting a representative sample. It is also equipped with a filtration system to remove particulate matter and sample coolers to cool the sample to a safe temperature. The sample container is latched into place, and the operator is completely out of harm's way when the sample is being caught. Any vapors from the sample are vented to a safe location and returned to the process.

However, suppose you or your teammate are called upon to collect a sample of a sour product, a product known to contain a high hydrogen sulfide concentration over 20 ppm. In that case, you must follow all site procedures and safety precautions. The following is a guide and can be used if your site does not have detailed safety precautions:

1. All personnel involved in this sampling must recognize that hydrogen sulfide is a deadly toxin and has killed many people in our industry. A brief encounter at high concentrations can leave a person paralyzed and can kill in minutes.
2. Personnel must wear proper PPE, including:
   - Impervious gloves

- Goggles (not required if wearing a full-face respirator)
- Chemical-resistant clothing

- Personnel should be wearing a calibrated and bump-tested H_2S detector with continuous monitoring capability and audible, visual, and vibration alarms.
- The breathing zone must be protected by wearing a full-face, positive-pressure, SCBA. The positive pressure prevents hydrogen sulfide leaks into the face mask.
- It is highly recommended that the person collecting the sample be outfitted in a positive-pressure SCBA and that a standby person also wearing an SCBA be located out of the immediate area but who has a clear view of the operation.
- Personnel should ensure that the work area is properly ventilated and should be working from the upwind side of the work area.
- Sample equipment should include plastic sample containers resistant to H_2S and hydrocarbons. Note that some metal containers can react with hydrogen sulfide and are not recommended for this purpose.
- Samples are best taken from a continuous flow through a three-way flow valve to ensure a representative sample. Otherwise, it may be necessary to overflow the container to ensure a representative sample. If the latter is used, be aware of the potential for H_2S release.
- When transporting the sample, ensure the containers are tight and not leaking. Never transport samples in the cab of a pickup truck or inside any vehicle. Transportation in a truck bed with ample ventilation is generally acceptable.
- All electrical tools and equipment in the area must be suitable for Class 1, Division 1 areas; in other words, they must be "explosion proof."

Climbing to the Top of a Tank Truck or Tank Car

Suppose you are climbing to the top of a tank truck or a tank car to open the hatch, possibly gauging the tank or collecting a sample, especially if the vessel is carrying oil well distillate, hydrocarbons or sour water, or any other product that may contain hydrogen sulfide. In that case, you should always have a hydrogen sulfide detector on your person. This is also true if the truck or rail car carries sour water, especially wastewater from sewage or sewer treatment plants, septic tanks, etc. More likely than not, this wastewater will have a high hydrogen sulfide concentration, and the vapors may be deadly.

This H_2S detector will provide you with an early warning of the presence of H_2S and allow you to exit the vessel safely. You can return properly outfitted with an SCBA and carry out the job safely. H_2S gas detectors are readily available from most safety suppliers, including Amazon. A good H_2S gas detector can be purchased for under $50, and they can save your life! This is extremely cheap insurance when lives are at risk.

Likewise, when climbing to the top of a rail car or tank truck, the vessel should be protected by fall protection. Many people have died when they were on the top of a tank truck or rail car and were overcome by hydrogen sulfide. This can happen easily when you or other workers are near an open tank truck or rail car top hatch. When the H_2S is at high concentration, it takes only one breath, and the worker can be knocked down; they are overcome, collapse, and fall to their death from the top of the vessel. Most often, their death was not from the H_2S, but it was due to the injuries sustained in the fall. This is where fall protection can save your life. This can be an improvised fall protection system using a harness connected to the top of the tank truck, or it can be specially engineered fall protection system with guardrails and a built-in tie-off for connecting the harness. Please see the links in the Additional Resources section of this chapter for more information on some of these engineered fall protection systems. There are other designs available on the web.

I use the same analogy for fall protection as the hydrogen sulfide detector. We should treat these like we do seat belts in a vehicle. As I have said before, when I get into a vehicle, it is an automatic response to buckle up; we should do the same for fall protection. Use that harness or ladder cage, just like the H_2S detector; if you are not using them, they cannot protect you in the event of a fall or if H_2S overcomes you. We must use these systems, or they cannot save you from a tragic fall and its potentially deadly consequences.

Figure 6.2, provided by Standfast Tram Fall Protection Systems, clearly shows what can happen when a worker falls from the top of a tank truck. This is much

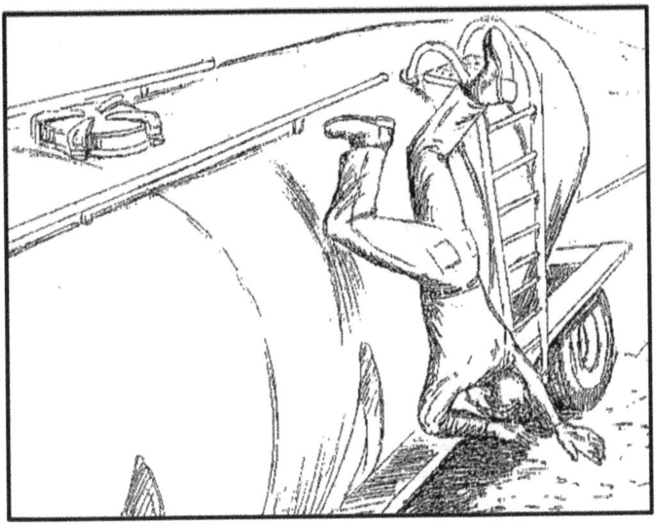

Figure 6.2 An illustration of a worker falling from the top of a tank truck. *Source:* Standfast TRAM Fall Protection System, Chicago, IL/Standfast, LLC.

more likely to occur when the worker is overcome by inhaling hydrogen sulfide vapors. One breath of highly concentrated hydrogen sulfide has a knockout effect, which can instantaneously render a worker unconscious, causing the worker to fall from the truck and sustain a serious injury or even death.

The following figures provide several designs for truck and rail fall protection systems; however, when you search the websites, you will find many others. These vary from self-standing ladders and platforms to various types of restraint systems. Some are manually operated, and several are self-powered. The main point is that whatever type of system you have, please be sure to use it. When working with and around hydrogen sulfide, don't leave it to chance – protect yourself and ensure that your workmates do the same.

This is not hypothetical. Please check Appendix 2: you will quickly find several cases in the OSHA documentation where this happened. Workers have been overcome by H_2S from an open hatch and have fallen, resulting in either severe injuries or death. When properly used, an engineered fall protection system can save the worker's life.

Figure 6.3a–c illustrates a worker accessing the top of a tank truck and using the Standfast TRAM Fall Protection System. When properly used, this type of

Courtesy of Standfast TRAM Fall Protection System, Chicago, IL

Figure 6.3 Illustrations of Standfast TRAM Fall Protection System available on their website. (a) The secure personnel anchor point and the harness. The anchor point in this design slides across the entire top of the tank truck. (b) A worker accessing the truck and the fall protection system. (c) A worker on top of the truck protected by the fall protection system. Other designs and models are available on their website; please see the link in the Additional Resources section of this chapter. https://www.sftram.com/tram?gad_source=1&gad_campaignid=21199838065&gbraid=0AAAAADNgUDIQovT3bkgOhb-c1m4MkIY-N&gclid=EAIaIQobChMI9puWmd-ojQMVzjbUAR20EQRZEAAYASAAEgK6lPD_BwE

Courtesy of Malta Dynamics, Malta, Ohio

Figure 6.4 Illustration of a portable manual-operated lifeline support system available from Malta Dynamics. Malta X550 is shown; other designs and models are available on their website. Please see the link in the Additional Resources section of this chapter.

Courtesy of Malta Dynamics, Malta, Ohio

Figure 6.5 Illustration of a portable lifeline support system available from Malta Dynamics. Other designs and models are available on their website. Please see the link in the Additional Resources section of this chapter: https://maltadynamics.com/mobile-fall-protection/

fall protection system will save workers' lives by preventing falls from the tops of tank trucks or rail cars.

The following photographs illustrate several other designs and types of fall protection and lifeline support systems. Figures 6.4 and 6.5 illustrate fall protections systems provided by Malta Dynamics and Figures 6.6 and 6.7 illustrate the systems provided by

Courtesy of Sam Carbis Solutions Group, LLC

Figure 6.6 Illustration of Carbis Access & Fall Protection, Safe Access to Tanker Truck is shown. Available from Sam Carbis Access & Fall Protection, Suppliers of Fall Protection Equipment; other designs and models are available on their website. Please see the link in the Additional Resources section of this chapter.

Courtesy of Sam Carbis Solutions Group, LLC

Figure 6.7 Another ahead of illustration of Carbis Access & Fall Protection, Safe Access to Tanker Truck is shown. Available from Sam Carbis Access & Fall Protection, Suppliers of Fall Protection Equipment; other designs and models are available on their website. Please see the link in the Additional Resources section of this chapter.

Sam Carbis Solutions Group. These fall protection systems will definitely save lives when used properly. These manufacturers are included in the list of Additional Resources at the end of this chapter, and others are readily available on the web.

Responding to a Release

If you are working on a process unit, a drilling rig, or other process facility that processes relatively high concentrations of hydrogen sulfide and a release is reported, this must be recognized as a very serious event and can easily lead to a fatality or even multiple fatalities. My guidance is to never venture into the area until you have properly donned an SCBA. With this done, you can determine what is being released, confirm if the release is flammable or toxic, and start the development of a response plan.

A very important consideration is the wind direction in every release. Never attempt to approach a release from downwind; this is common sense, but it is worth a reminder. We have had fatalities due to personnel panicking and running into the scenario from downwind. Always approach from crosswind and, if possible, move upwind away from the release after the release has been identified. When notifying other responders, this is valuable response information, and you will want to pass the wind direction along. Again, if you suspect the release contains hydrogen sulfide, ensure that other responders are aware and equipped with an SCBA. If the release is flammable, the standard approach is to begin water deluge as soon as possible. Afterward, the primary strategy should be to isolate the source of the release by blocking the failed equipment. Always isolate remotely if possible.

What Do Ranchers and Farmers Have in Common with a Sewerage Plant Worker?

Hydrogen sulfide is also generated by the biological breakdown of organic waste such as decaying cattle feed, hay, and especially manure from cattle, equine, or swine. This is especially true in areas with low oxygen content or anaerobic conditions that can occur in pits or bogs or the bottom of silos or bins containing last year's feed waste or manure waste. In these areas, hydrogen sulfide can not only be generated but may also be present at extremely high concentrations. Any disturbance of the waste can result in the release of hydrogen sulfide, and the vapors have killed ranchers and stablemen alike.

The same natural biodegradation processes exist at municipal or industrial sewer treatment plants. Human waste is treated with biological microorganisms to convert the waste matter into safe discharge into our streams and rivers. This activity is under anaerobic or very low oxygen conditions and generates significant amounts of hydrogen sulfide. This activity starts early in the sewer system and is typically present in sewer piping, sewer transfer pumping stations, and other sewerage equipment upstream of the waste treatment plant. For example, we live in the suburbs, and there is no sewer treatment, and we use a Mo-Dad Wastewater Treatment System, which provides effective treatment. However, occasionally, when cutting the yard near the Mo-Dad vent, I can smell the rotten egg odor of H_2S.

Today, sewer plants and ranchers are being exposed to hydrogen sulfide, some unexpectedly, and unfortunately, some are killed by this exposure. Appendix 2 of this book contains many case studies of exactly this happening.

From my perspective, there is a general need for education among this group of workers on the extreme hazards of hydrogen sulfide. Learning the hazards and how these incidents have happened in the past is an effective way to help prevent future incidents. However, it is not that simple. Generally, these groups don't have access to the same analytical tools, such as H_2S detection equipment, that are provided to industrial workers. I ask you, when is the last time that you saw a cattleman or rancher with an H_2S detector, or a municipal sewer line worker with a detection device. While I recognize some municipalities do use these, I believe there is a lot of work to do in this area.

By way of this textbook, I appeal to the trade unions and waste treatment industry groups to reach out to their constituents to help fill this gap.

Use of Self-contained Breathing Apparatus

If personnel are expected to work near vessel openings or to open pipe flanges anywhere where the presence of hydrogen sulfide may be greater than 20 ppm, or other inert or toxic gases may be present, consideration should be given to requiring the use of an SCBA. We should also require a standby person in SCBA. See Figure 6.8 for an example of an SCBA.

SCBAs are typically available in negative-pressure or positive-pressure configurations and several breathing air tank sizes and pressures. Generally, the higher the tank pressure, the longer the response or rescue breathing time.

I prefer the positive-pressure regulator in combination with a full-face mask. The positive pressure will supply air at pressure to the mask anytime the pressure difference is reduced. The pressure inside the mask is slightly above ambient pressure, ensuring positive airflow is always available to the user. This can overcome minor leaks in the mask, preventing the leakage of smoke or vapors into the mask.

The negative-pressure system is quickly becoming obsolete; however, it may still be available in some areas. The disadvantage of the negative-pressure regulator is that the user activates it by creating negative pressure inside the mask when breathing. In the event of leakage around the face mask, the negative-pressure regulator may allow toxic vapors or smoke into the mask.

Courtesy of MSA Safety

Figure 6.8 Example of a self-contained breathing apparatus with a full-face mask.

Practice Makes Perfect

I remember that when I was an operator, we frequently practiced getting into our SCBAs when we had a quiet weekend. It became a game, and we practiced to see how quickly we could put the SCBA on. The objective was to get it on and fully ready to use in under 30 seconds. Of course, as I said, this became like a game, but it was great practice, and we were all ready to get into an SCBA within seconds. I suspect this is still happening today; however, we shouldn't take it for granted. Please ensure that your operators know the location of the SCBAs in their work areas and that they are familiar with how to use them in an emergency.

Emergency Plan

When working around high concentrations of hydrogen sulfide or if sampling high concentrations of hydrogen sulfide, always have an emergency plan and ready availability for medical assistance if needed. For example, if an unexpected release occurs, immediately move all personnel upwind of the release and notify the emergency response team.

Hydrogen sulfide is a deadly gas and can kill with short-term exposure. Always be aware of the OSHA short-term exposure limits of 20 ppm ceiling and 50 ppm peak.

SCBA Breathing Air

SCBAs are readily available from almost all industrial safety suppliers, and when used properly, they can save lives. Anytime a company implements a requirement for employees to use respiratory protection, they must have a written Respiratory Protection Program, and the program must meet the requirements as spelled out in OSHA Standard 29 CFR 1910.134. This standard requires the following: a written Respiratory Protection Plan, the user must be medically cleared to use the respirator, the employee must be trained, and they must be fit tested,

One consideration is that the breathing air used for SCBAs must be industrial grade D. The stringent breathing air requirements are established by the US Occupational Safety and Health Administration (OSHA), and detailed requirements are outlined in ANSI/CGA Commodity Specification for Air, G-7.1-1989. This includes specific limits for oxygen, hydrocarbons, carbon monoxide, and carbon dioxide and a lack of noticeable odor. These limits are as follows:

- Oxygen content percentage by volume: between 19.5 and 23.5%
- Hydrocarbon (oil and particulate): maximum of 5 mg per cubic meter of air
- Carbon monoxide: maximum of 10 ppm by volume
- Carbon dioxide: maximum of 1,000 ppm by volume
- Lack of noticeable odor

Due to several breathing air cylinders in the industry being shipped, which were filled with gas other than breathing air, it is a good idea to have procedures and requirements in place to conduct a quality check with a gas test instrument on all breathing air cylinders received from an offsite supplier. Of course, these quality checks should be completed before the cylinders are available for use in the field.

Remember, oxygen should never be used as breathing air. Breathing pure oxygen, especially for long periods of time, can be harmful, resulting in lung damage, oxygen toxicity, and other health issues. This is why oxygen should only be administered under the observation and care of medical professionals.

Chapter 6: Key Lessons Learned

- It's important to recognize that a release of high concentration of hydrogen sulfide, generally 1,000 ppm or higher, can kill instantaneously.
  - In these concentrations, the victim is knocked down with the first breath, and they must be revived within 2–4 minutes; otherwise, severe brain damage or death will occur.
- All personnel, when in the field where H_2S is being processed or is present, even in moderate concentrations (concentrations of 20 ppm or more), should wear an H_2S detector that has been bump-tested against a known standard to ensure that it is fully functional.

  - Anytime the detector goes into alarm status, everyone in the area should immediately evacuate and return to investigate only after properly donning a SCBA (note that 10 ppm is the NIOSH-REL).
- Anytime personnel are in the process of opening process equipment for insertion of blinds, blind removal, opening access plates, or any other act of OPE (opening process equipment), the vessel or piping system must be verified that there is no pressure or substantial product present (zero energy).
  - It should also be tested to confirm that it is safe to open. This means that the vessel or piping system is gas tested and confirmed at or below 10% of the LEL, 20 ppm or less of H_2S, and 19.5% O_2 or more.
- Each point of opening for blind insertion, blind removal, access plate removal, or any other vessel or piping system opening must be clearly identified by the process organization.
  - Generally, this is done by hanging a wearproof or weatherproof tag with a unique design, which is only used for this purpose on the flange to be opened.
  - The flanges to be opened should also be marked at each opening point on the P&ID drawing.
- Several manufacturers make available a closed-loop sample system designed to lock a sample container and collect a representative sample while keeping the operator completely out of harm's way.
- When sampling streams containing H_2S, personnel must wear proper PPE, including:
  - Impervious gloves
  - Goggles (not required if wearing a full-face respirator)
  - Chemical-resistant clothing
  - Personnel should wear a calibrated, bump-tested H_2S detector with continuous monitoring capability and audible, visual, and vibration alarms.
  - The breathing zone must be protected by wearing a full-face, positive-pressure, SCBA. The positive pressure prevents hydrogen sulfide leaks into the face mask.
  - It is highly recommended that the person collecting the sample be outfitted in a positive-pressure SCBA and a standby person also wearing an SCBA is located out of the immediate area, but who has a clear view of the operation.
- When responding to a release, a very important consideration is the wind direction in every release. Never attempt to approach a release from downwind.
- Hydrogen sulfide is also generated by the biological breakdown of organic waste such as decaying cattle feed, hay, and especially manure from cattle, equine, or swine.
  - This is especially true in areas with low oxygen content or anaerobic conditions that can occur in pits or bogs or the bottom of silos or bins containing last year's feed waste or manure waste.
- Human waste is treated with biological microorganisms to convert the waste matter into safe discharge into our streams and rivers.

  - This activity is under anaerobic or very low oxygen conditions and results in the generation of significant amounts of hydrogen sulfide.
- When working around high concentrations of hydrogen sulfide or if sampling high concentrations of hydrogen sulfide, always have an emergency plan and ready availability to medical assistance if needed.
  - For example, if an unexpected release occurs, immediately move all personnel upwind of the release and notify the emergency response team.
- Anytime personnel climb to the top of a rail car or tank truck that may contain a product containing hydrogen sulfide such as oil well distillate or sour water, they should be wearing an H_2S detector, and the vessel should be equipped with fall protection.

Hydrogen Sulfide Flammability

Hydrogen sulfide is extremely flammable and has a very low flash point and a wide flammable range. It is easily ignited by open flames, arcing, or sparking electrical equipment. Before issuing a hot work permit in areas where H_2S may be present, always conduct a gas test for flammables (LEL) and hydrogen sulfide. All electrically powered tools and equipment used in these areas must be rated for Class 1, Division 1 electrical classified areas.

Chapter 6: Incident(s) Related to the Key Lessons Learned

Worker Fatality Resulting from High Concentration of Hydrogen Sulfide in a Sewer Manhole, a Loss of Consciousness and a 30-foot Fall (1 Fatality)

On July 3, 2023, an employee climbed down a fixed ladder within a sewer manhole structure to retrieve a stuck tool caught on a ladder rung. As the employee began to climb out of the manhole opening, the employee lost consciousness and fell approximately 30 feet to the bottom of the manhole. The employee died as a result of toxic, fatal levels of hydrogen sulfide gas exposure and blunt force injuries to the head resulting from the fall.

Death of a Worker While Pumping Wastewater from Sludge Containment Area Due to Inhalation of Hydrogen Sulfide

At 9:00 p.m. on September 19, 2022, an employee was pumping wastewater out of a drain sludge containment area when, after the operation was complete, he lifted the suction hose and fell to the ground. The employee died due to acute thiosulfate intoxication from inhalation of hydrogen sulfide.

Employee #1 Died After Chemical Exposure. Employees #2, #3, #4 Attempted Rescue, Employee #2 also Lost Consciousness. Caused Thought to be Exposure to Hydrogen Sulfide. Employees #2, #3, and #4 Were Treated and Released

At 7:00 a.m. on August 22, 2022, Employee #1 was operating an excavator at a hazardous waste treatment and disposal facility. Employee #1 was working in the Solids Processing Unit and was following a batch sheet recipe to mix neutralizing agents with industrial waste. Employee #1 went to the Liquids Processing Unit and asked a coworker to send spent sulfuric acid "pickle liquor" to the bay where he was working. Employee #1 returned to his excavator so he could mix the batch when the "pickle liquor" was sent. Moments after the "pickle liquor" was sent, Employee #1's excavator swung into the sidewall of the Solids Processing Unit. Employee #2 rushed over to see what happened and found Employee #1 unconscious. When Employee #2 attempted to climb up into the excavator to help Employee #1, he became lightheaded and lost consciousness. However, Employee #2 was able to regain consciousness and go for help and 9-1-1 was contacted. Additional coworkers arrived to assist Employee #1, including Employees #3 and #4. Employee #1 was transported to the hospital and treated for respiratory chemical exposure, but was declared brain dead on August 24, 2022, and died the following day. Employees #2, #3 and #4 were observed at the hospital and released the same day with no diagnosed injuries. It is unknown what chemical the employees were exposed to but was suspected to be hydrogen sulfide.

Additional Resources

Industrial Safety Products, BW Technologies BWC2-H Clip Single Gas H_2S Monitor, https://www.industrialsafetyproducts.com/bw-technologies-bwc2-h-clip-single-gas-h2s-monitor/?sku=BWC2-H-1&srsltid=AfmBOooQOQOMVZa4I2Aye2qxvHrUlTTmiAxcc1BO56rDSX0cRb1x8d8z3Lw&gPromoCode=ISP50-Q41&gQT=1

H2S Monitor | USA NIST Calibration, https://www.forensicsdetectors.com/products/hydrogen-sulfide-detector?variant=30700793528364&country=US&currency=USD&utm_medium=product_sync&utm_source=google&utm_content=sag_organic&utm_campaign=sag_organic&srsltid=AfmBOoodDeYlcAeI_G6dcR4_ihaSP53azS-RgIVnLRrFaQ8NPQjvEBmJ4rY&gQT=1

Swagelok® Grab Sampling Systems, https://www.swagelok.com/en/blog/closed-loop-sampling-reduce-emissions

DOPAK® PRODUCT OVERVIEW, Crane Instrumentation & Sampling, https://dopak.com/wp-content/uploads/Product_Overview_Crane_Dopak_0123.pdf

Modcon Systems Ltd., CLOSED LOOP SAMPLE SYSTEMS, https://www.modcon-systems.com/wp-content/uploads/2014/05/Sampling-Systems-en.pdf

Crane Sample Systems, Manual Continuous Sampler Systems, https://texassampling.com/manual-continuous-sample-systems/

MSA, The Safety Company, Self-Contained Breathing Apparatus (SCBA), https://us.msasafety.com/Supplied-Air-Respirators-%28SCBA%29/Self-Contained-Breathing-Apparatus-%28SCBA%29/c/11701?content_index_en_us%5Bquery%5D=&product_index_us_en%5Bconfigure%5D%5BhitsPerPage%5D=20&product_index_us_en%5Bconfigure%5D%5BclickAnalytics%5D=true&product_index_us_en%5Bconfigure%5D%5BfromCategoryListing%5D=true

US Occupational Safety and Health Administration (OSHA), Standard Interpretations: Clarifications for breathing air quality and use, https://www.osha.gov/laws-regs/standardinterpretations/2023-05-31

iSi, Grade D Breathing Air: What Requirements Do I Have to Meet for My Breathing Air System?, https://isienvironmental.com/grade-d-breathing-air-blog/

Standfast Tram Fall Protection Systems, 4809 N. Ravenswood Ave, Unit 315; Chicago, IL 60640 USA, https://www.sftram.com/?_vsrefdom=GAW&gad_source=1&gad_campaignid=18389649089&gbraid=0AAAADNgUDJpelOCKPrgZ0e4uhWS8iDbY&gclid=EAIaIQobChMItuOt3NGHjgMVxwxECB241Q8eEAAYASAAEgILK_D_BwE

Sam Carbis Solutions Group, LLC, Florence County, South Carolina, https://www.carbissolutions.com/solutions/?utm_source=google&utm_medium=ppc&utm_campaign=multiview_google&gad_source=1&gclid=EAIaIQobChMI2fvX6fbAjAMVXS9ECB1O3x7MEAAYASAAEgIBuPD_BwE&gclsrc=aw.ds https://www.carbissolutions.com/solution/liquid-trucks/

Malta Dynamics (Information on Malta X550, and other models), info@maltadynamics.com, 210 13th Street, Malta OH, 43758, https://landing.maltadynamics.com/xseries?utm_source=google&utm_medium=cpc&utm_campaign=xseries&utm_content=ad_group_1&utm_term=fall%20protection%20solution&utm_term=fall%20protection%20solution&utm_campaign=XSERIES&utm_source=adwords&utm_medium=ppc&hsa_acc=3130872631&hsa_cam=20429519659&hsa_grp=151418414785&hsa_ad=668308109457&hsa_src=g&hsa_tgt=kwd-407788635726&hsa_kw=fall%20protection%20solution&hsa_mt=p&hsa_net=adwords&hsa_ver=3&gad_source=1&gclid=EAIaIQobChMI2fvX6fbAjAMVXS9ECB1O3x7MEAAYBCAAEgIIkvD_BwE

Chapter 6: End-of-Chapter Quiz

1. When you think about the hazards of hydrogen sulfide and its toxicity, H_2S is primarily an __________ hazard.
   Please select the most correct answer:
   a. toxic
   b. skin and eye irritation
   c. inhalation
   d. neurological

2. A release of high-concentration hydrogen sulfide, generally 1,000 ppm or higher, can kill __________. In these concentrations, the victim is knocked down with the first ________ and they must be revived within ___ to ___ minutes.

3. All personnel, when in the field where ________ ________ is being processed or is present, even in moderate concentrations (concentrations of 20 ppm or more), should wear a H_2S ________ that has been bump-tested against a known ________ to ensure that it is fully functional.

4. When personnel are in the process of opening process equipment for insertion of blinds, blind removal, opening access plates, or any other act of OPE (opening process equipment), the vessel or piping system must be verified that there is no ________ or substantial ________ present (________ energy).

5. Each point of opening for blind insertion, blind removal, access plate removal, or any other vessel or piping system opening must be clearly identified by the ________ or operations organization. Generally, this is done by hanging a ________ or ________ tag with a ________ design and only used for this purpose on the flange to be opened.

6. When sampling streams containing H_2S, personnel must wear proper personal protective equipment (PPE) including the following
   Please list as many of the PPE items as you can:

7. What is one of the most important considerations when responding to a release potentially containing hydrogen sulfide?
   Please choose the most important consideration:
   a. The atmospheric temperature and humidity
   b. The wind direction
   c. The process pressure at the time of the release
   d. Ensuring that we can keep the unit in service

8. When working around high concentrations of hydrogen sulfide or if sampling high concentrations of hydrogen sulfide, always have an emergency plan and ready availability to medical assistance if needed.
   a. For example, if an unexpected release occurs, immediately move all personnel ________ of the release and notify the ________ ________ ________.

9. When responding to a release, never venture into the area until you have properly donned a ____-________ ________ ________. With this done, you can determine what is being released, confirm if the release is flammable or toxic, and start development of a response plan.

10. What is the primary benefit of using an engineered closed-loop sample system when sampling toxic products, for example, products containing high concentrations of hydrogen sulfide?
    Select the most appropriate answer:
    a. An engineered sample system ensures that the sample is representative of the process.
    b. This type of system saves costs as the sample containers are reusable.
    c. Product losses are minimized with an engineered system.
    d. The operator is out of harm's way when the sample is being caught.

7

What Has Happened/What Can Happen Again?

Introduction

In this chapter, I will take you through several significant incidents that have occurred and, unfortunately, are still occurring due to exposure to the toxic effects of hydrogen sulfide (H_2S). As you will see in Appendix 2, so many of these have occurred, and the Appendix only reflects those incidents in the United States. From my experience, incidents like these are happening worldwide. Unfortunately, we don't have a good database of these incidents worldwide and the lessons learned from each one.

Possibly the most tragic industrial incident involving H_2S occurred in 2004 when, on the evening of December 23, a gas well in Kaixian County, Southwest China's Chongqing Municipality, blew out, releasing tons of highly toxic hydrogen sulfide gas into the community. The gas blowout quickly spiraled out of control when the blowout preventer (BOP) valve failed to function. The release resulted in the deaths of 243 people. Most of these deaths occurred while they were sleeping or were attempting to flee from the deadly release. Additionally, over the next few days, 4,000 others were hospitalized due to their injuries, and more than 60,000 were displaced from their homes and communities. Many cattle and other animals, including wildlife, were also killed due to the toxic effects of the release.

According to local media, an investigation of this deadly release of H_2S concluded that the primary cause was workplace negligence by the Eastern Sichuan Drilling Company operating under the China National Petroleum Corporation (CNPC). The issue was that the drilling company had failed to repair a broken well blowout prevention valve, which could have stopped the release if it had been functional. Also, emergency procedures broke down. The gas well should have been ignited immediately to prevent the spread of the toxic gas. Unfortunately, this was not done for about 18 hours, long after most of the harm had been done.

Hydrogen Sulfide (H_2S) Hazards: Risk Recognition and Prevention, First Edition. M. Darryl Yoes.

Contributing to the disaster in 2004, communications in this remote region were less than adequate, and it was difficult to keep the residents informed of the hazards and actions they should take to protect themselves.

Of course, this was 20 years ago, and conditions have vastly improved. Technology has advanced, and we have learned much more about process safety and hence the current requirement for Process Safety Management Systems, including Permit to Work Standard, Confined Space Entry Standard, Operations Procedures, Process Hazard Analysis, Management of Change, Control of Defeat of Safety Critical Devices, Prestart up Reviews, and many other process safety work processes. We also have very effective process control systems with advances in distributed control systems, programable logic controllers (PLCs), etc. However, as you see in this book, we still have a way to go as releases and deaths due to H_2S exposure are still occurring, as evidenced by the October 2024 release and fatalities at the Pemex Refinery in Deer Park, Texas. This is the first case study we will review, although the Chemical Safety Board is currently investigating, and it will be a little time before we fully understand what happened there.

Another very serious incident was reported by the National Library of Medicine and National Center for Biotechnology Information (NIH), where six workers were killed while attempting to unblock a wastewater cistern. Reportedly, the first worker collapsed and died shortly after clearing the obstruction, and the other five entered the cistern, one after the other, to help their colleagues. Autopsies confirmed that all six workers died due to poisoning from hydrogen sulfide. The NIH reported that hydrogen sulfide (H_2S) is one of the most serious toxic gases encountered in forensic practice. Aside from being a by-product of many industrial processes, this gas is naturally produced during the decomposition of organic substances.

As stated before, unfortunately, this type of loss is not uncommon. It is a very natural and human urge to immediately attempt rescue when we see one of our workmates collapse inside a confined space or in a process area where toxic materials like hydrogen sulfide may be present. Often, the would-be rescuer collapses directly on top of the person they are attempting to rescue, which delays the ability of emergency responders to reach the first person who collapsed. This comes down to the training of all who are on the scene. This human urge must be resisted, and training must be allowed to kick in. The bottom line is that if you are not trained in rescue techniques and do not have immediate access to specialized PPE, such as a self-contained breathing apparatus (SCBA) and proper extrication equipment, you must call for the emergency responders and standby to lead them to the site. Otherwise, the data suggests you may also become a victim and another statistic. Please see the link to the NIH report in the Additional Resources section of this chapter for additional information on the incident mentioned here.

Several local and international news agencies reported a similar accident in Bastrop County, Texas, just SW of Austin, Texas, in August 2023. In that incident, three hunters from Florida had been invited down to Texas to hunt hogs. According to news reports and the Bastrop County Sheriff's Office, they were running their dogs when one of the dogs accidentally fell into an old water cistern. A companion hunter who lived in the local area ran to his truck to call 911 for help. However, it was obvious that another hunter quickly jumped into the cistern to rescue the dog and became incapacitated. It was also painfully obvious that he was quickly followed into the cistern by the remaining two hunters who were now attempting to rescue him. It was also obvious that the last two hunters who entered the cistern prepared to enter by partially removing their clothes before jumping in.

When emergency responders arrived, there was no sign of life. The toxic hydrogen sulfide vapors in the cistern killed the three hunters and the dog. Recovery efforts were delayed as the H_2S concentration was very high, the water level was well below the small opening into the cistern, and the area around the cistern was unstable. It required first responders from several agencies working for several hours to extract the victims from the cistern. It is noteworthy that water cisterns of this type were reported to be commonplace in the county and that, according to the local sheriff, state law requires that a cover be installed over the cistern. This cistern had no such cover.

The root causes for incidents 8 and 9 are very simple and, in some ways, very similar. I seriously debated whether I should include these; however, they are included since they show how quickly H_2S can strike if we do not follow procedures and wear proper protective equipment. In both cases, this is what happened, and in both cases, it resulted in death. This is a hard price for failure to follow procedures or not taking the time to don the proper safety gear. As you review the following case studies, please be aware of what the operator was doing at the time and of their surroundings. Please try to put yourself in their situation and think about the hazards and how they were addressed.

Incident 1: October 10, 2024; Pemex Refinery, Deer Park, Texas. Hydrogen Sulfide Release When Workers Were Attempting to Install Isolation Blinds (2 Fatalities, 35 Injured)

Mexican state-owned petroleum company Pemex
Deer Park, Texas
Incident Date: October 10, 2024 (2 Fatalities, 35 Injured, 13 Hospitalized)

This incident reportedly happened as contractors were "working on a flange," according to Pemex, the Mexican state-owned oil company. This was a very large release of highly toxic hydrogen sulfide gas (H_2S) and resulted in what was reported as the instantaneous deaths of 2 of the contract workers and 35 other injuries, including the hospitalization of 13. The release was reported to be almost 20 tons of hydrogen sulfide and about 14 tons of sulfur dioxide. The release reportedly lasted several hours and disrupted activities in two nearby cities. The communities of Deer Park and portions of Pasadena received orders to shelter in place. Also, sections of nearby motorways were closed until the release subsided, and the area was again deemed safe.

Pemex released a press release, which included the following:

"This event involved the release of hydrogen sulfide from the Amine Regeneration Unit at the refinery, which lasted only a few minutes. The release of gases occurred during the execution of maintenance activities in this process unit, and containment tasks began immediately.

The refinery team blocked the line and significantly reduced emissions in about an hour. The release of gases from the process equipment was mitigated and stopped in approximately three hours."

The following is from the latest release by the US Chemical Safety and Hazard Investigation Board website shortly after they released a statement that they are investigating this incident: "10 October 2024, at approximately 4:30 p.m., a hydrogen sulfide release occurred at the PEMEX Deer Park facility's amine recovery unit while workers were attempting to remove isolation blind from piping that normally transports hydrogen sulfide gas. The incident caused two fatalities and 13 injuries."

The CSB provided another important update on their investigation status on November 20, 2024, stating, "On the day of the incident, two contract workers at the refinery partially opened a flanged connection on piping containing hydrogen sulfide gas, which caused the release. One of these workers was fatally injured. The work was supposed to be done on a different, isolated piping segment about five feet away from the opened flange. Two other contract workers from a separate company, who were working on equipment less than 250 feet away and downwind from the release, were also affected, with one fatally injured from the hydrogen sulfide exposure."

This is an important update because now we know that the wrong pipeline was opened, one that had not been prepared for opening by the operations group and had not been isolated, depressured, and verified as zero energy. The CSB investigation is continuing, and we will get the full details when it is complete and released. However, I encourage each reader to read Chapter 6 of this book, "How the Hazards of Hydrogen Sulfide Are Controlled," as this scenario is discussed in more detail, including how this kind of incident can be prevented at your facility.

The CSB issued another update on March 19, 2025, which included the following details on the incident: "On the day of the incident, maintenance contractors were working to removing (sic) piping isolation devices, called blinds, from ARU6, one of the refinery's Amine Regeneration Units (ARUs). During the task, workers inadvertently opened a flange on a piping segment of another unit, ARU7, which was still pressurized with hydrogen sulfide gas. At approximately 4:23 p.m., the ARU7 piping flange was opened, releasing a toxic concentration of hydrogen sulfide gas into the air." We are now certain of what piping was opened and that this piping was in super-high concentration H_2S service, operating with approximately 90% H_2S (900,000 ppm), operating at full pressure and operating conditions.

In a petroleum refinery, the Amine Unit uses amine, typically monoethanolamine (MEA) or methyl diethanolamine (MDEA), to recover the hydrogen sulfide (H_2S) and carbon dioxide (CO_2) from hydrocarbon streams. This is done in two process vessels: an amine scrubber and an amine regenerator. The hydrocarbon stream, typically naphtha product from hydrotreater or hydrocracker units, is circulated in contact with the amine solvent in the scrubber vessel, where the solvent absorbs the H_2S and CO_2 from the process stream. The H_2S-rich amine then goes to the amine regenerator, where the H_2S gas is contacted by steam. This drives the H_2S-rich vapor overhead from the regenerator and is processed further in the sulfur plants, where it is converted to elemental sulfur.

Notes from the CSB Update on this incident dated November 2024:

- On October 10, 2024, at 4:23 p.m., two contract workers partially opened a flanged connection on the Amine Recovery Unit (ARU7) piping where hydrogen sulfide was flowing (Figure 7.2), resulting in the release of hydrogen sulfide gas and fatally injuring one of the workers. Figure 7.1 shows the large response to the incident scene due to the massive release of hydrogen sulfide.
- The depressurized, purged, and isolated ARU6 piping segment, which was the location of the permitted work, was approximately five feet from the ARU7 piping flange that was opened.

Figure 7.1 Photograph of emergency response teams to Pemex Refinery during hydrogen sulfide release on October 10, 2024.

Figure 7.2 Photograph indicating the flange that was out of service and permitted to open and the in-service flange that was opened.

Notes from the CSB Update on this incident dated November 2024:

- Two contract workers from a separate company, unrelated to the ARU task, were working on equipment less than 250 feet away and downwind from the release (Figure 7.3). One of these workers was also fatally injured from the hydrogen sulfide release.
- The release continued for approximately one hour until PEMEX Deer Park Refinery operators and emergency responders isolated the leak by closing upstream valves. The PEMEX Deer Park Refinery reported that approximately 27,000 pounds of hydrogen sulfide were released into the atmosphere.

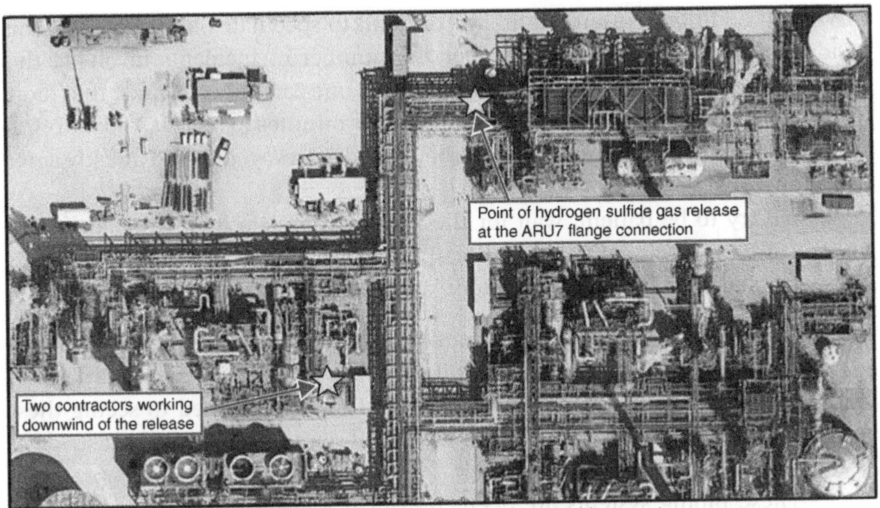

Courtesy of US Chemical Safety and Hazard Investigation Board (CSB)/Public Domain

Figure 7.3 Photograph indicating the area at the Pemex Refinery where the hydrogen sulfide release occurred, and contractors were working downwind of the release.

- Thirteen other contract workers were transported to nearby hospitals for medical evaluation. The neighboring cities of Deer Park, Texas, and Pasadena, Texas, issued shelter-in-place orders that lasted several hours.

The gas from these two process vessels is extremely rich in H_2S in both cases. For example, the gas from the amine regenerator may be as high as 80–90% H_2S or higher. This is equivalent to 800,000–900,000 ppm H_2S. The CSB reported that, in this case, the concentration of the released H_2S was 90% (900,000 ppm). When you realize that the NIOSH IDLH (National Institute for Safety and Health Immediately Dangerous to Life and Health) for H_2S is only 100 ppm, one can appreciate the concentration of this gas and its deadly nature. From my perspective, when this is considered, the press release by Pemex appears to downplay the seriousness of this release and its impact on the workers and the local community.

The US Chemical Safety and Hazard Investigation Board (CSB) immediately dispatched a team of investigators to the site and investigated the incident. The Chairperson for the CSB, Steve Owens, released the following statement: "This is a very serious incident that caused multiple fatalities and injuries and potentially put the surrounding community at risk."

The CSB will certainly release a detailed report on this incident, including the cause, causal factors, and recommendations to help prevent a recurrence of this deadly event. The CSB often develops video animations and great safety videos that can be used in safety training programs. I use many CSB videos in the Safe Operations Training I lead for petroleum and petrochemical facilities in the United States and worldwide.

Without additional information, it is difficult to speculate on the cause of this incident. However, when you consider the number of incidents involving this deadly gas and the frequency of these incidents, one can only conclude that more work must be done. I think that if you take a few minutes to review Appendix 2, a long list of these incidents just here in the United States, you would quickly agree that more must be done.

We will try to be patient, wait for the CSB to complete its investigation, and understand this incident in more detail. However, based on what we already know coupled with our incident history, we can do the following immediately:

- Ensure that all equipment containing high concentrations of H_2S is well identified in the field and on all unit drawings (Process Instrumentation and Flow Diagrams or P&IDs). For example, I am accustomed to seeing these high-concentration H_2S pipelines identified with a unique colored stripe or colored bands on the piping.

  These piping systems are deemed critical and receive priority for safety reviews and consideration when work is planned on or near them. This is extremely important when planning work in the vicinity of these pipelines, for example, other equipment opening or lifting operations in or near the area where these pipelines are located.
- The location of each blind point or equipment opening is identified, tagged, or otherwise identified in the field and the unit drawings. This should be done with a unique, weatherproof, tamperproof tag designed to be used only for opening process equipment (OPE). It should be installed on each flange before issuing the OPE permit. This unique tag is an indication that operations have prepared the equipment, which is at a zero-energy state and is ready for opening.
- Each flange or access plate to be opened or removed should also be identified on the unit P&ID drawings before the work permit to open is issued.
- Ensure that when we are opening equipment near the equipment containing high concentrations of H_2S, zero energy is always confirmed before breaking any flanges or opening any process equipment. A proper bleeder unplugging device will be used if a vent or drain is fouled. All OPE work will cease immediately, and flanges will be retightened if any leakage or spillage occurs.
- When opening piping or equipment that has been deemed critical – for example, it is in high concentration of H_2S service or is located near piping or vessels in critical service – there should be a requirement for a process representative, typically an operator, to stand by for each opening into the equipment. This is to ensure that the correct flanges are being opened, that all bolts are not removed before the flange is opened, and that we truly do have a zero-energy state. Not removing all the bolts means that the flange should be loosened when opening, allowing the craftsman to quickly reclose the flange if it is pressurized or contains product.

- All personnel, when in the field on units containing or processing gas or liquids with high concentrations of H_2S, should always be equipped with a personal H_2S detector. These devices should be thought of as wearing seat belts in a vehicle. They can save your life and should always be worn around equipment containing high concentrations of H_2S.
- All personnel performing equipment opening or maintenance on flanges on high H_2S piping systems or equipment should wear either a hose-line supplied breathing air mask or an SCBA pack.
- When preparing a process vessel for entry, it must be empty of its contents and purged with a positive-pressure air purge to remove any trace flammables and other contaminants. The air purge is typically done using an electrically powered fan or blower and air ducts to direct the fresh air to remote sections of the vessel. If the work is being done where flammables or hydrogen sulfide may be present or in an electrically rated area, the device must be rated for a Class 1, Division 1 atmosphere.
- Before entering any process vessel, such as a fractionation column, a storage tank, a tank truck or tank car, or any other kind of confined space entry, one should always ensure the vessel has been tested for oxygen (O2), flammables (lower explosive limit [LEL]), H_2S (hydrogen sulfide), and other toxics specific to the equipment or process.
- All confined space procedures must be in place before entering a process vessel, including a gas test, a standby person located at the vessel entry who communicates with the entrant(s) and can summon rescue if needed, and a personal rescue plan.

  These and other requirements associated with confined space entry are spelled out in the OHSA Permit Required Confined Spaces regulation (29 CFR 1910.146).

The CSB Final Report will be available on the CSB website (csb.gov) once the investigation is completed. I encourage all readers to obtain a copy of this report and become familiar with this incident. Also, ensure that the recommendations are implemented at your facility.

OSHA Control of Hazardous Energy Standard (29 CFR 1910.147)

The OSHA standard for Control of Hazardous Energy (29 CFR 1910.147) covers the servicing and maintenance of machines and equipment in which the unexpected energization or start-up of the machines or equipment or release of stored energy could cause injury to employees. This standard establishes minimum performance requirements for the control of such hazardous energy.

The standard requires employers to establish a program and utilize procedures for affixing appropriate lockout or tagout devices to energy-isolating devices and disabling machines or equipment otherwise to prevent unexpected energization, start-up or release of stored energy, and injury to employees. The standard requires lockout or tagout, which shall be used and supplemented by the procedural and training requirements of the standard.

The standard provides more detail, but the objective is to ensure that all energy sources have been determined before the work permit is issued and that potential energy sources have been isolated, locked out, and tagged out. In cases of blinding of process lines, all connecting lines must be identified, depressured, and drained, and all interconnecting lines must be locked out and/or tagged out by the provisions of the standard. The standard details the types of lockout/tagout devices and how these devices should be identified in the field.

The overall objective of this standard is to ensure that all systems are energy-free and that interconnecting lines have been identified and isolated to ensure that the system being opened is energy-free.

Incident 1: Key Lessons Learned

October 10, 2024: Pemex Refinery, Deer Park, Texas: Hydrogen Sulfide Release

This incident remains under investigation by the US CSB, and there is still much to learn. However, what has been reported thus far indicates that work was being done on a flange in the amine recovery unit. The wrong flange was opened, releasing approximately 27,000 pounds of super-rich concentrated hydrogen sulfide into the atmosphere. This incident resulted in the deaths of 2 contract employees and a total of 35 additional injuries, of which 13 were reported to be hospitalized. The amine systems contain piping and other equipment in super-rich hydrogen sulfide service, including concentrations as high as 80–90% (800,000–900,000 ppm) or higher, whereas the OSHA maximum exposure is 20 ppm.

The piping and other super-rich hydrogen sulfide service equipment should be designated as critical systems, and special attention should be required during any inspection or maintenance work on or near these piping systems. The piping systems should be identified with labels or a painting system to identify them as critical systems containing the super-rich H_2S. A special walk-down must be carried out whenever work permits are issued for work on or near these critical systems.

A special or unique weatherproof and tamperproof tag should be posted on each critical system flange to indicate that the flange has been prepared for opening by the operations group and that the operations group has verified it to be at a zero-energy state. To ensure the correct lines are being opened, a representative of the operations group should be present for each flange or bleed valve opening on these systems, especially for those sections of pipe in high-concentration hydrogen sulfide or nested among piping containing high-concentration hydrogen sulfide where the worker can easily get off onto the wrong flange and inadvertently open a section of pipe that is still pressurized with high concentration of H_2S.

The workers assigned to open the flanges must understand that they can only open those tagged with the unique "line opening" tag. No other flanges are to be opened under the issued line opening work permit.

Workers should never remove all the bolts from a flange until the flange is cracked open and has been proven to have zero energy. Removing all the bolts can leave the flange "stuck" in the closed position, and if it is pressurized, a massive release can occur when it is opened. Cracking the flange open will allow the worker to quickly reclose it if it is found to be pressurized or if the product leaks from the flange. Please refer to Chapter 6 "How the Hazards of Hydrogen Sulfide Are Controlled" for additional information.

Additional Resources

Reuters, Hours-long toxic gas leak at Pemex oil refinery near Houston far exceeded legal limit, https://www.reuters.com/business/energy/pemexs-deer-park-oil-refinery-scales-back-operations-after-fatal-accident-2024-10-12/

USA Today, 2 dead, 35 injured after chemical leak of hydrogen sulfide at Pemex Deer Park oil refinery, https://www.usatoday.com/story/news/nation/2024/10/11/chemical-leak-texas-pemex-deer-park/75624727007/

US Chemical Safety and Hazard Investigation Board (CSB), CSB Launches Investigation into Pemex Hydrogen Sulfide Release (November 10, 2024), https://www.csb.gov/csb-launches-investigation-into-hydrogen-sulfide-release/

US Chemical Safety and Hazard Investigation Board (CSB), Update on Pemex Hydrogen Sulfide Release (November 20, 2024), https://www.csb.gov/us-chemical-safety-board-releases-investigation-update-into-fatal-hydrogen-sulfide-release-at-pemex-deer-park-refinery-in-deer-park-texas/

National Library of Medicine, National Center for Biotechnology Information, Domino effect: An unusual case of six fatal hydrogen sulfide poisonings in quick succession, https://pubmed.ncbi.nlm.nih.gov/26869537/

Incident 2: Hydrogen Sulfide Release at Aghorn Operating Waterflood Station; Aghorn Operating Inc. (2 Fatalities – 1 Company Fatality, 1 Civilian Fatality)

Investigated by the US Chemical Safety and Hazard Investigation Board (CSB)
Incident Date: October 26, 2019 (Two Fatalities – An Operator and His Wife Were Both Killed in This Incident)

This incident was investigated by the US CSB, and a detailed report was released with recommendations to prevent a recurrence of this tragic event. It was also briefly mentioned in the Note to Readers in this book. A summary of the CSB Final Report is provided here, and a link to the CSB Final Report is provided in the Additional Resources section.

The Aghorn Water Flood Station is located near Odessa, Texas, in the Permian Basin and is an integral part of the Foster Field, which comprises 68 oil and gas wells. The water flood station accumulates sour water from the producing wells and reinjects the water back into the formation to help maintain the pressure in the gas and oil underground wells. The station consists of sour water settling and holding tanks and three positive-pressure motor-driven reciprocating pumps used to inject the water into the underground formations. The water is separated from recovered oil by settling in the tanks before being reinjected into the formation. The tanks are in open areas within a fenced-in area, and the injection pumps are inside an enclosed "pump house." See Figures 7.4 and 7.5 for photographs of the Aghorn Waterflood Station.

The water and hydrocarbons are rich in hydrogen sulfide (H_2S). Therefore, the pump house is equipped with H_2S alarms and ventilation fans to help clear any accumulation of H_2S from the pump house in the event of a spill or leak. The building also has large bay doors, typically between 50 and 75% open. The CSB could not confirm if the ventilation fans were operating at the time of the incident. They did confirm that the H_2S alarms were not functional and were not working at the time of the incident.

There are three positive displacement pumps in the pump house; typically, two are in service, and the third is set up as a spare. A small, separate, and enclosed control room in one corner of the pump house housed a computer and a PLC that controlled the operation of the pumps. The operator used a touchpad to interface with the PLC to adjust operational parameters, alarm set points, etc. In the event of a pump issue, the PLC sends a phone notification to the on-duty operator (referred to as the pumper), and the pumper drives to the pump house to address the reason for the alert.

Courtesy of US Chemical Safety and Hazard Investigation Board (CSB)/Public Domain

Figure 7.4 Photograph of the Aghorn Waterflood Station storage tanks, pump house, and control room building.

Courtesy of US Chemical Safety and Hazard Investigation Board (CSB)/Public Domain

Figure 7.5 Overhead view of the Aghorn Waterflood Station storage tanks, pump house, and control room building.

Two pumpers are assigned to the Aghorn facility: Pumper B, the lead pumper, and Pumper A, the relief pumper. Pumper B was on vacation and out of town when this incident occurred. Pumper A was the lead pumper on the day of the incident.

On the evening of the incident (October 26, 2019), and while at home, Pumper A received a computer notification of a trouble alarm on one of the positive displacement pumps. He left his wife and two children at home and drove to the pump house to check on the status of the sour water injection pumps.

The CSB concluded, based on evidence at the scene, that the pumper was in the process of isolating one of the operating pumps when he was overcome by H_2S and succumbed to the toxic gas. They found that the discharge valve was about 95% closed and the suction valve was about 50% closed. They also noted that one of the plungers for this pump was broken, and sour water was released into the pump house onto the pump house floor through the opening in the broken plunger. Please see Figures 7.7 and 7.8 for photographs of the failed positive-pressure water injection pump.

Notes from the CSB Final Report:

Figure 7.6 illustrates the Aghorn oil extraction and waterflooding processes.

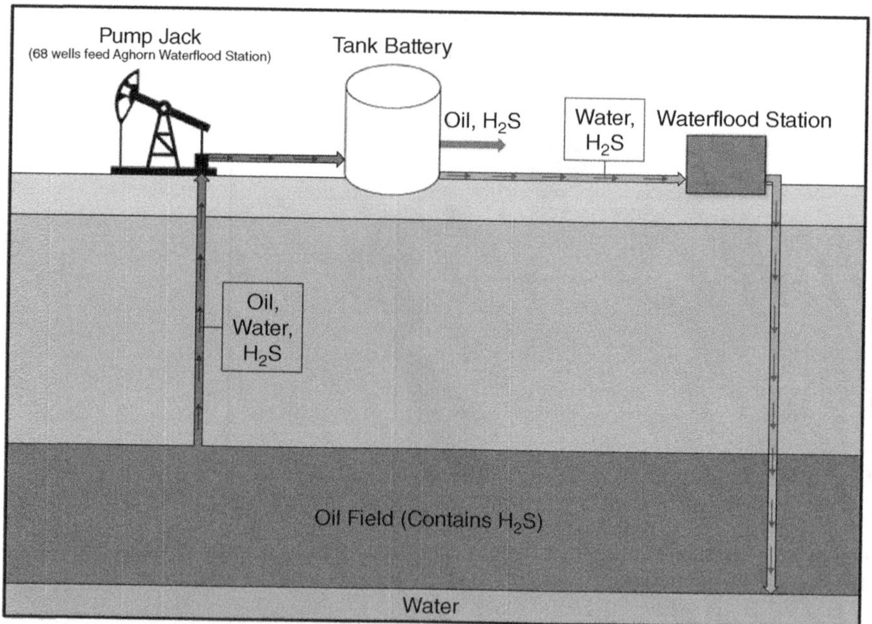

Courtesy of US Chemical Safety and Hazard Investigation Board (CSB)/Public Domain

Figure 7.6 Simplified illustration of the Aghorn waterflooding process from the CSB Report.

Figure 7.7 Photograph of the failed positive pressure reciprocating sour water injection pump. Note that sour water was released from the failed (unthreaded) plunger into the confined area of the pump house.

- Large pumps, called pump jacks, extract oil from underground oil reservoirs and transfer the oil to a tank battery through pipes. The oil is stored in large tanks at the tank battery, where produced water separates from the oil.
- Produced water typically contains other components, including hydrogen sulfide (H_2S), a toxic gas in the area's oil and gas reservoirs.
- After separating it from the oil, other pipes transfer the produced water from several tank batteries to a waterflood station.
- The waterflood station pumps the produced water back into the oil reservoir.

The wife of Pumper A became concerned about her husband after not hearing from him for several hours. She attempted to call her husband several times and received no response. She then called Pumper B, who was on vacation and out of town. He explained that he was most likely in the pump house. The wife, now worried, put two of her small children, ages 6 and 9, in her car and drove to the facility; she found the gate unlocked and opened, drove into the facility, and parked in front of the pump house.

After more time passed, Pumper A was also concerned and notified an Aghorn foreman, requesting him to notify 911 and have someone check the facility. The 911 dispatcher quickly sent an ambulance from Odessa, and the paramedics smelled very strong H_2S as soon as they arrived. They decided to back out of the area and, while doing so, noticed the two small children in the wife's car. They quickly drove the ambulance to a safe location, set up a command center, and one of the paramedics donned an SCBA and walked to the wife's car, where he found the children still conscious in the car. He drove the car back to the command center, where the children would be out of harm's way.

Courtesy of US Chemical Safety and Hazard Investigation Board (CSB)/Public Domain

Figure 7.8 A close-up photograph of the failed positive-pressure reciprocating pump, clearly showing the failed plunger and the source of the sour water release.

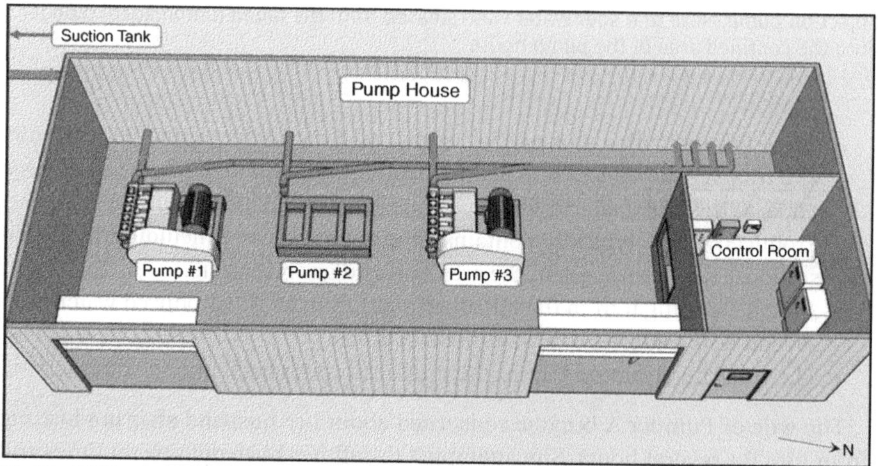

Courtesy: US Chemical Safety and Hazard Investigation Board (CSB)/Public Domain

Figure 7.9 Overview of the Aghorn Station Pump House, highlighting the pump locations and Control Room.

The responders found Pumper A and his wife in very close proximity to each other, and both were deceased. The responders measured H_2S concentration at one location at 150 ppm, and in other locations, the readings exceeded the design of the detection equipment.

Notes from the CSB Final Report:

Figure 7.9 illustrates the Aghorn Waterflood Station Pump House layout.

- The waterflood station also has an H_2S detection and alarm system. If activated, a red, rotating beacon light will illuminate on top of the pump house, and an automatic phone notification system will engage and notify the on-duty pumper of the H_2S release.

- The pump house has large bay doors, exhaust fans on the west wall opposite the bay doors, and natural vents on the four outside walls. The bay doors and exhaust fans were the pump house's primary means of ventilation.
  - Aghorn did not have a procedure to ventilate the pump house. Through interviews with an employee and written communication with the company, CSB investigators were informed that these overhead doors were typically kept between 50 and 75 percent open.
  - The CSB was unable to confirm whether the exhaust fans were operational at the time of the incident.

As other Aghorn employees arrived, they found the failed pump still running and sour water pouring out of its cradle. Aghorn employees quickly isolated all the wells connected to the Aghorn site.

The responders located Pumper A's personal hydrogen sulfide detector in his pickup truck. The detector was in alarm when found, indicating that it had detected dangerous concentrations of hydrogen sulfide, even though it was outside the building where the release occurred. This is an important element in this tragic event and must be emphasized. If the Pumper had been wearing the detector, he would have realized that he was entering an extremely dangerous environment, and it is possible that his life and the life of his wife would have been saved. Wearing the H_2S detector should be a mandatory employment requirement and must be enforced by management. I will talk more about this as we close out this chapter.

The Chemical Safety Board issued a detailed and rigorous report on this tragic incident, including the causal factors and recommendations to prevent a recurrence.

Causal Factors (as identified by the CSB):

- Failure to wear the personal hydrogen sulfide detection device. More importantly, the CSB found that the facility management did not require wearing a personal detection device. From my perspective, these are life-saving devices, and their use should be considered a mandatory condition of employment and enforced.
- LOTO (Lockout/Tagout) was not used by Pumper A to de-energize the failed pump before attempting to isolate it. The CSB also found no written policy or procedure requiring LOTO to be used. The CSB also concluded that more than likely, the auto-start function of the PLC started the pump, even though it had failed. The pump should have been locked out to prevent it from starting, including electrically isolated. Comprehensive Lockout/Tagout is required by OSHA 1910.147, The Control of Hazardous Energy (Lockout/Tagout).

- Pump house Confinement was the third critical safety issue identified by the CSB. The building was an enclosed structure with ventilation by natural vents on each of the four walls, coupled with the practice of leaving the large bay doors about 60% open, the position they were in at the time of this incident. There was also ventilation provided, although there was no evidence the fans were being used. The CSB concluded that building confinement and lack of forced ventilation contributed to the very high levels of H_2S on the evening of the incident.
- The fourth key safety issue identified was the lack of a formalized Safety Management Program. A formalized Process Safety Management System should be the cornerstone of process safety management, which did not exist at Aghorn. I discuss this in detail in the textbook "*Process Safety Management: The What, Why, and How Behind Safe Petrochemical Plant Operations.*"
- The CSB also found that the H_2S Detection and Alarm System was nonfunctional during this incident. The pump house was designed and equipped with H_2S detection to send signals from the detection elements to the PLC. However, the system was not functional and failed to trigger either of the two H_2S alarms.
- Site security was also found to be less than adequate. The informal policy was to leave the entrance gate to the facility open and unattended when an Aghorn employee was working in the facility. This policy resulted in the gate being open, allowing the wife of Pumper A to drive directly into the facility and enter the pump house.

Recommendations to prevent recurrence (an abbreviated summary of the CSB recommendations – not all are included here):

- Personal H_2S monitors should be provided for all employees and visitors at all waterflood stations where there is the potential to expose personnel to concentration of 10 ppm or higher.
- All Aghorn facilities: Develop a site-specific, formalized, and comprehensive lockout/tagout program, including policies and training, as detailed in OSHA 29 CFR 41910.147. This should include energy control procedures and periodic inspections to ensure compliance.
- All Aghorn facilities: Where the potential exists to expose workers or visitors to 10 ppm H_2S or higher, conduct an independent and comprehensive analysis of each facility's design and mitigation systems to ensure ventilation in the event of an accidental release.
- All Aghorn facilities: With the potential to expose workers or visitors to 10 ppm H_2S or higher, develop and demonstrate the use of a Safety

Management System focused on protecting workers and non-employees from H_2S. This includes risk identification, assessment, mitigation, and monitoring of design, procedures, maintenance and training related to H_2S. This must comply with OSHA 29 CFR 1910.1000 (Air Contaminants) and 29 CFR 1910.146 (The Control of Hazardous Energy – Lockout/Tagout).

- All Aghorn facilities: Where the potential exists to expose workers or visitors within the perimeter of the facility to 10 ppm H_2S or higher, ensure that the H_2S detection and alarm systems are properly maintained and configured and develop site-specific procedures based on manufacturer specifications, current codes, standards, and industry good practice guidance. This must address installation, calibration, inspection, maintenance, training and routine operations.
- All Aghorn facilities: Where the potential exists to expose workers or visitors within the perimeter of the facility to 10 ppm. H_2S or higher, develop and implement a formal, written, site-specific security program to prevent the unknown and unplanned entrance of those not employed by Aghorn, starting with a requirement for employees to lock access gates upon entering and departing the facility.
- To the Occupational Safety and Health Administration (OSHA): Issue a safety information product (such as a safety bulletin or safety alert) that addresses the requirements for protecting workers from hazardous air contaminants and from hazardous energy.
- To The Railroad Commission of Texas (The commission that regulates the oil and gas industry (and several other industries) in the state of Texas): Develop and send a Notice to Operators to all oil and gas operators that fall under the jurisdiction of the Railroad Commission of Texas that describes the safety issues described in this report, including:
  1. Nonuse of Personal H_2S Detector
  2. Nonperformance of Lockout/Tagout
  3. Confinement of H_2S Inside Pump House
  4. Lack of Safety Management Program
  5. Nonfunctioning H_2S Detection and Alarm System
  6. Deficient Site Security
- The CSB also included the Key Lessons for Industry in their report, which essentially includes the essence of these Recommendations. Please see the Final Report for more details.

This was a tragic incident that resulted in the deaths of an oil field worker and his wife due to the toxic nature of hydrogen sulfide. We should all strive to improve safety when working with or near streams rich in hydrogen sulfide so that their lives are not lost in vain. This could have been prevented if the recommendations above had been fully implemented.

It is noteworthy that, according to the Associated Press, the US Justice Department issued a statement that a federal grand jury in Midland, Texas, has charged Aghorn Operating Inc. and its Vice President with violating federal clean air laws and alleging that they obstructed a federal job safety investigation.

The Railroad Commission of Texas, Oil and Gas Division, the regulatory authority for these facilities in Texas, sent a "Notice to Operators" informing all operators of this incident and the importance of the CSB recommendations and how they relate to the Texas Administrative Codes (TAC).

Incident 2: Key Lessons Learned (Author's Comments)

October 26, 2019: Aghorn Operating Waterflood Station, Odessa, Texas, in the Permian Basin (2 Fatalities)

This was a particularly tragic incident in that a plant operator and his wife were both killed by exposure to toxic hydrogen sulfide. The operator responded to a pump alarm after normal work hours, and his wife went to check on him when he failed to return home. Both were found dead lying near a sour water injection pump that had failed and was leaking sour water, water containing high concentrations of H_2S. What is particularly troubling is that the operator's H_2S detector was found in his vehicle and not on his person after the incident.

The US CSB investigated this incident and discovered several additional deficiencies in process safety management at this facility, including nonfunctioning H_2S alarms, insufficient forced air ventilation in a building containing equipment with high concentrations of H_2S, and inadequate security for the perimeter to control access.

I am especially troubled by the operator leaving his H_2S detector in his vehicle when he entered the building to check on the pump alarm. The detector was reported to be in alarm status, indicating it had detected H_2S. I think this is a case where the functioning detector could have, and probably would have, saved his life and, subsequently, the life of his wife as well. I liken wearing of an H_2S detector to the use of a seat belt when riding in a vehicle. When I get into a vehicle, I certainly don't expect to be involved in a major crash; however, each time, I automatically buckle my seat belt without thinking about it. I watch my family, my wife, my kids, and my grandkids, and they do the same thing automatically every single time they enter a vehicle. This is setting expectations and then following up with enforcement. We must get to the place where operators and others do the same thing with their gas test detectors when they enter a process area. They should automatically, without thinking about it, ensure that they have the detector on their person and that it is functional.

Ultimately, I see this responsibility should be placed on company management and policies. I think there should be a policy that wearing H_2S detectors is a condition of employment. If any employee or contractor violates this policy, they are putting their job on the line. I know this sounds harsh, but wearing detectors anytime you are in the presence of high-concentration hydrogen sulfide is that important. These detectors will save lives if this policy is put in place. However, a policy must be implemented, communicated to all employees and contractors, and continuously enforced for this to work. Otherwise, it is worthless, and lives will continue to be lost.

I can't help but think about the situation that was discovered by the first responders when they arrived at the facility. Upon arrival, the children were discovered in the wife's vehicle and were unharmed. Upon discovering the children, their vehicle was immediately relocated to the command center, which was well away from the pump house, to get them to a safe location. It is so fortunate that they remained in the vehicle and didn't attempt to follow their mother into the pump house.

The pump house bay doors were approximately 60% open when this incident occurred, and the first responders found a very strong odor of H_2S near the pump house. They immediately decided to respond by wearing an SCBA. The pump house was designed with hydrogen sulfide detectors that were designed to illuminate a flashing red light on the top of the pump house if H_2S was detected. The CSB confirmed that none of the responders noticed the red light as being operational at the time of the incident. The detectors were later tested by the CSB and failed to function properly when they were tested. Also, Aghorn could not produce documentation of periodic testing of the H_2S detectors to ensure their operability.

Of course, the responders were responding wearing SCBA, but when they entered the pump house, it must have been a horrible sight. Produced sour water containing significant quantities of hydrogen sulfide was leaking from the broken pump plunger onto the floor of the pump house. H_2S readings on the responders' detectors ranged from 150 ppm to off-the-scale high. The operator was obviously in the process of isolating the suction and discharge valves to the pump when he was overcome by the H_2S. His body was found lying adjacent to the failed sour water transfer pump. His wife was also discovered very close to her husband, and she was also dead, having also been overcome by the hydrogen sulfide vapors.

There are many lessons to be learned from this incident. I encourage each reader of this book to open the link to the US CSB and read the Aghorn report in detail. This incident and these people's lives are for naught if we don't learn from it and take action to ensure that it doesn't repeat at one of our facilities.

Additional Resources

US Chemical Safety and Hazard Investigation Board (CSB), Aghorn Operating Waterflood Station Hydrogen Sulfide Release, https://www.csb.gov/aghorn-operating-waterflood-station-hydrogen-sulfide-release-/

US Department of Justice, Office of Public Affairs, Charges Filed in Connection with Texas Oilfield Deaths, https://www.justice.gov/opa/pr/charges-filed-connection-texas-oilfield-deaths

RAILROAD COMMISSION OF TEXAS, Oil and Gas Division, NOTICE TO OPERATORS, U.S. Chemical Safety Board Recommendations Following, their Report on the 2019 Aghorn H_2S Incident, https://www.rrc.texas.gov/media/kycavdof/nto_csb-2019-aghorn-h2s-incident-report_2-6-2023.pdf

US Occupational Safety and Health Administration (OSHA), Hydrogen Sulfide, Evaluating and Controlling Exposure, https://www.osha.gov/hydrogen-sulfide/evaluating-controlling-exposure

Energy Safety Canada, CONTROLLING CHEMICAL HAZARDS, Guidance Sheet, HYDROGEN SULPHIDE (H_2S) GAS MONITORS, https://www.energysafetycanada.com/EnergySafetyCanada/media/ESC/Resources/Guidance%20Sheets%20-%20Important%20HSE%20Topics/Hydrogen_Sulphide_H2S_Gas_Monitors.pdf

Industrial Scientific, Understanding H_2S Gas: A Silent Hazard in Industrial Environments, https://www.indsci.com/en/blog/understanding-h2s-gas-silent-hazard-industrial-environment

Incident 3: Accidental Transfer of Sodium Hydrogen Sulfide (NaHS) into a Tank Containing Dilute Sulfuric Acid, National Transportation Safety Board (1 Fatality, 1 Injury)

US National Transportation Safety Board – Washington, D.C. 20594
Hazardous Materials Accident Brief:
Whitehall Leather Company tannery, Whitehall, Michigan
Incident date: June 4, 1999 (1 Fatality and 1 Injury)

The following is based on work completed by the US CSB. The CSB has published a safety bulletin on the hazards of sodium hydrogen sulfide (NaHS). This safety bulletin is referenced in the Additional Resources section with a link where it can be accessed. Sodium hydrogen sulfide is also known by the following chemical names:

- Sodium bisulfide
- Sodium hydrogen sulfide
- Sodium hydrosulfide
- Sodium hydrosulfide solution
- Sodium hydrosulfide [solution]
- Sodium hydrosulfide solution
- Sodium sulfhydrate

Emergency response codes:

- CAS No.: 16721-80-5
- UN No. (solution): 2922 (Figure 2)
- DOT Hazard Class: 8 (Figure 2)
- EPA Hazardous Waste No.: D003

The CSB reported that since 1971, reported incidents involving liquid solutions of sodium hydrosulfide (NaHS) have resulted in 32 deaths and 176 injuries, most notably in the leather tanning and pulp and paper industries.

The most serious safety concern associated with NaHS is its capacity to produce large amounts of deadly hydrogen sulfide gas (H_2S) when it reacts with an acid or is exposed to high heat.

Warning:

H_2S hazard: Ensure separation of NaHS and acids. Allowing spilled NaHS to inadvertently mix with an acid ($pH < 7$) poses a significant risk of harm. Similarly, if a sewer system is not designed to control H_2S, allowing NaHS to run off to a sewer containing acids also poses a significant risk. Large amounts of toxic and flammable H_2S may evolve very rapidly! Please see the CSB Safety Bulletin for additional information.

The June 1999 incident involving one fatality and one injury, which occurred at the Whitehall Leather company, was investigated by the US National Transportation Safety Board. This is a summary of this investigation. This report provides a link to the complete investigation in the Additional Resources section.

Whitehall Leather is a leather tanning facility and uses ferrous sulfate, sulfuric acid, and sodium hydrosulfide in the tanning process. These products are stored and then used from separate storage tanks on the property. All three chemical storage tanks and off-loading connections are in different areas at the facility. The company receives all three of these products routinely delivered by tank truck. The ferrous sulfate transfer area was on the northeast side of the facility and away from the other transfer areas.

It was routine to receive ferrous sulfate truck deliveries during all shifts. Generally, the Environmental Manager assists the driver set up to off-load the product. However, when a truck arrives during the second or third shift, it is customary for the shift supervisor or possibly an operator to meet the driver and help set up the delivery. The company representative would meet the driver and escort them into the facility, showing them the off-loading connection and the compressed air connection used to pressurize the truck for the off-load.

Following the off-load, the company representative typically signed the invoice verifying that the truck cargo had been received. These procedures were not in writing, and there was no formalized training for these off-load procedures.

A cargo truck driver arrived at the Whitehall Leather Company at about 3:30 a.m. on the morning of June 4, 1999. The driver had never delivered to this site and required assistance to locate the delivery point and unload his product, a truckload of sodium hydrosulfide solution. Upon arrival, he requested assistance from a tannery employee who called the shift supervisor who had just completed unloading a load of "pickle acid." Pickle acid is a low pH acid product, typically a weak solution of citric or muriatic acid. The shift supervisor assumed this was another load of pickle acid and walked the driver to the storage tanks for pickle acid. The shipping papers on the current cargo clearly stated that this was a load of sodium hydrosulfide. However, the shift supervisor neither asked for the papers nor verified the shipping documents. Please see Figure 7.10 for a general overview of the Whitehall Leather Company facility, including the three chemical storage tanks and the offloading connections.

The shift supervisor escorted the driver to the ferrous sulfate connection, the only working transfer connection at that location. He then unlocked the gate and allowed the driver to drive his vehicle to the offloading area designated by the shift supervisor. The shift supervisor was asked to sign the shipping documents, which

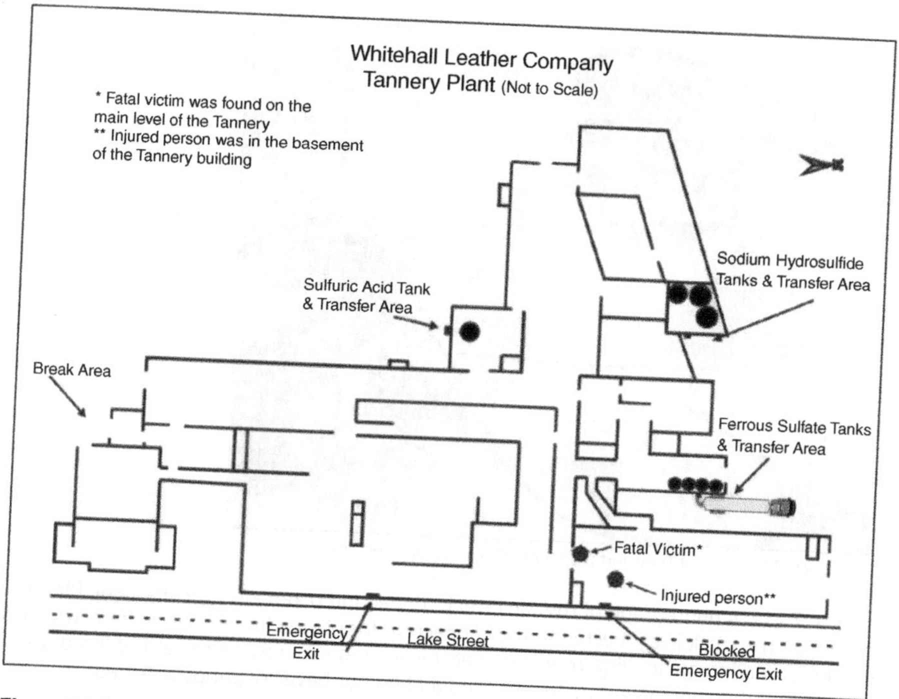

Figure 7.10 A general overview of the Whitehall Leather Company facility indicating the location of the three chemical storage tanks and the offloading connections. Also shown here is where the two victims were following the accident.
Source: The US Transportation Safety Board/National Transportation Safety Board. Public Domain.

he did without reading the documents. He then left the area. The papers he signed included the following statement: "I have checked the documents for this shipment and verified that there is adequate storage room to receive this shipment, and connection has been made to the proper storage facility."

There were no plant employees in the product transfer area, and the driver found the off-loading hose already connected to the off-loading connection on the side of the building, which was labeled "FERROUS SULFATE." After the incident, the opposite end of this same hose was discovered to be connected to the connection on the truck used for off-loading products. The investigators determined that sodium hydrosulfide solution had been transferred to the tank containing ferrous sulfate, allowing mixing of the two noncompatible chemicals. The chemical mixing of sodium hydrosulfide and ferrous sulfate produced a chemical reaction that released copious amounts of hydrogen sulfide, a poisonous gas. Figure 7.11 illustrates the hose from the tank truck connected to the labeled "Ferrous Sulfate." A close-up of this same photograph is shown in Figure 7.12.

Courtesy of The National Transportation Safety Board/Public Domain

Figure 7.11 A photograph of the tank truck connection taken after the accident. Please note that the truck is connected to the connection for off-loading ferrous sulfate, which is the only working off-loading connection at this location. The two pipes at the left of the ferrous sulfate coupling are no longer used and do not have couplings attached.

Courtesy of The National Transportation Safety Board/Public Domain

Figure 7.12 A close-up photograph of the tank truck connection attached to the off-loading connection labeled "Ferrous Sulfate." Sodium hydrosulfide and ferrous sulfate are incompatible; mixing them will result in toxic hydrogen sulfide gas formation. This photograph was taken by the US Department of Transportation and was included in the incident report.

After the release, the driver was found unconscious about 230 feet away, inside the tannery building. He was overcome and collapsed due to the effects of the toxic hydrogen sulfide gas. Notably, the investigation found no telephone or other means of communication near the transfer area. If this had been provided, the driver may have been able to notify plant personnel of the emergency. The accident investigators also found that the emergency valve at the rear of the cargo tank and the compressed air valve inside the tannery building, approximately 40 feet from the transfer area, were closed and secured. It is obvious that the driver realized what was happening and was attempting to stop the release.

Investigation findings (this is a summary; please see the US DOT report for the complete listing of findings and recommendations):

- The generation of H_2S and the resulting fatality was caused by the off-loading of sodium hydrosulfide into a storage tank containing ferrous sulfate. H_2S was generated by a chemical reaction between the two incompatible chemicals.
- The procedures for off-loading chemicals into the storage tanks were not documented.
- The company employees were not properly trained in the procedures for off-loading chemicals.
- The truck driver had been provided with a copy of a driver's manual during his last refresher training. This manual required the driver to follow the instructions of the company representative and to have the representative sign the release of responsibility form on the delivery receipt before connecting the truck to the storage tank.
- The company's representative did not verify the shipping papers to ensure the correct chemical product was being off-loaded.
- The company representative was not present when the connection was made to the truck and when the product transfer was started.
- There was no means for the driver to quickly communicate with the company representative in the event of an emergency.
- This company's unloading areas for sulfuric acid and sodium hydrosulfide were both equipped with padlocks to help prevent the unauthorized unloading of a chemical into a storage tank.
  - The ferrous sulfate off-loading connection was not locked, and no other protective devices were used to prevent accidental connections of other products.
  - Though not directly related to the accident, the investigation revealed that the padlock that was supposed to secure the sodium hydrosulfide transfer coupler was not installed but was lying nearby. Its condition indicated that it had been there for some time.

Additional Considerations:

- Procedures where a serious loss of containment can occur or which have the potential for employee injury or impact on the environment or community should be designated as "critical procedures."
- All procedures designated as "critical" should be in a checklist format, with the operator signing each step to indicate that it has been completed.
  - Critical procedures require a detailed Management of Change process for any revisions or changes.
- It is a generally accepted industry practice that in addition to the coupling being locked, the off-loading connections for incompatible chemicals should be color-coded and have a unique coupling for each chemical. This is to help ensure that accidental mixing of chemicals during off-loading is less likely to occur.

Three Other Incidents Similar to This:

Numerous incidents similar to what happened here have occurred in the industry. A very similar incident occurred at the MGPI Processing facility in Atchison, Kansas, on October 21, 2016, resulting in 140 injuries. Several people were admitted to the hospital, and 11,000 residents were either evacuated to a safe location or sheltered in place due to the toxic chlorine gas drifting across the area.

The Chemical Engineer, in an article discussing the MGPI chlorine release and the new safety video released by the CSB, reported that since January 1, 2014, eight other similar incidents have caused 44 injuries and the evacuation of 846 people. Please refer to the article and the CSB report on the MGPI release for more details. Links to both are provided in the Additional Resources section. Figure 7.13 illustrates the massive release of chlorine gas which occurred at the MGPI Processing facility due to the inadvertent mixing of chemicals during a truck off loading.

Courtesy of US Chemical Safety and Hazard Investigation Board (CSB)/Public Domain

Figure 7.13 Chlorine release from MGPI Processing Inc. in Atchison, Kansas, when sulfuric acid was inadvertently unloaded into a sodium hypochlorite tank.

October 2016

MGPI Processing, Inc.

Atchison, Kansas

A truck driver connected a truckload of sulfuric acid (H_2SO_4) and began unloading it into a sodium hypochlorite (NaClO) tank. The mixing of these incompatible chemicals resulted in a large chlorine gas cloud that resulted in 140 injuries, with a number of people admitted to hospital and 11,000 residents forced to evacuate or shelter in place. This incident was investigated by the US CSB, and a detailed report is available on the CSB website.

The cause of this incident was that the truck discharged sulfuric acid into a tank containing an incompatible chemical, sodium hypochlorite (better known in a less concentrated form as bleach).

This incident clearly illustrates what can happen when two incompatible chemicals are mixed. The CSB report provides information on how this incident occurred and great recommendations for a recurrence of this event. Please see the link to the CSB report in the Additional Resources section of this chapter.

Another Comparable Incident in Virginia

April 2014

Mecklenburg, Virginia

A similar incident occurred in Mecklenburg, VA, when a truck driver offloaded sodium hypochlorite (NaClO) into a tank containing hydrofluosilicic acid (H_2SiF_6).

In this incident, the chemical reaction released a cloud of chlorine gas, sending 11 people, including several firefighters, to the hospital.

Georgia-Pacific Naheola Mill

Pennington, Alabama (2 Fatalities, 8 Injuries)

16 January 2002

This incident was investigated by the US CSB, and a detailed report was issued. This is a summary of that report. A link to the CSB Final Report is included in the Additional Resources section of this chapter. You are encouraged to review the report and recommendations to prevent a reoccurrence.

The CSB classified this event as a "reactive chemical event," which the CSB defines as "a sudden event involving an uncontrolled chemical reaction with significant increases in temperature, pressure, or gas evolution that has caused, or has the potential to cause, serious harm to people, property, or the environment."

Courtesy of US Chemical Safety and Hazard Investigation Board (CSB)/Public Domain

Figure 7.14 Georgia Pacific Naheola Mill Collection drain next to tank truck unloading station.

On January 16, 2002, several contractors were working near a sewer manway that contained residual sodium hydrosulfide (NaHS) from prior truck unloading. Figure 7.14 shows the location of a hydrogen sulfide release at the Georgia Pacific Naheola Mill, adjacent to an underground drain system. On the day of the incident, sulfuric acid (H_2SO_4) was added to the acid sewer to control pH downstream in the effluent area. Sodium hydrosulfide (NaHS) from the oil pit and the collection drain flowed into the sewer and reacted with the sulfuric acid to form hydrogen sulfide (H_2S). Two contractors near the manway were killed by hydrogen sulfide poisoning; seven other contractors were injured by the H_2S exposure.

Incident 3: Key Lessons Learned

June 4, 1999: Whitehall Leather Company Tannery, Whitehall, Michigan (1 Fatality and 1 Injury)

This incident is where a chemical, a liquid solution of sodium hydrosulfide (NaHS), was transferred from a chemical delivery truck into a tank of ferrous sulfate ($FeSO_4$). These two chemicals are incompatible, and the mixing of the two results in the formation of hydrogen sulfide vapors. Unfortunately, there have been other incidents very similar to this in our industries where many people have either died or were seriously injured due to the toxic gases emitted by the mixing of incompatible chemicals.

The CSB reported that in about the past 50 years, incidents involving liquid solutions of sodium hydrosulfide (NaHS) have resulted in 32 deaths and 176 injuries, most notably in the leather tanning and pulp and paper industries. They also report that the most serious safety concern associated with NaHS is its capacity to produce large amounts of deadly hydrogen sulfide gas (H_2S) when it reacts with acid or is exposed to high heat.

In this case, the person receiving the cargo failed to verify the shipping documents, which clearly correctly identified the chemical; however, without checking the shipping papers, he assumed that the load was the same as the one he had received earlier. This resulted in the chemical being offloaded into the wrong tank. The company representative also escorted the truck driver to the wrong unloading connection and showed the driver the line to the ferrous sulfate tank instead of the sodium hydrosulfide tank. The company representative also signed the shipping papers without reading them, or he would have realized the connection was made to the incorrect tank.

This is a case where documented procedures for offloading chemicals should be required and written in a checklist format. Each step in the process is documented and signed by the operator as the step is completed. This includes verifying the shipping papers for the tank being unloaded to ensure the chemical names are properly matched. All unloading lines should be clearly labeled, as they were in this case. Also, the hose connections for each chemical should be noncompatible to help ensure that connections are not cross-connected between chemicals.

This incident involved the transfer of sodium hydrosulfide into the wrong tank, and there have been many of these incidents, with each one generating an uncontrolled release of hydrogen sulfide. Unfortunately, there have been many similar incidents like this one between other incompatible chemicals, each leading to a massive release of a different toxic gas into the atmosphere. A couple of additional incidents are included in this chapter to help communicate these hazards and lessons learned to help prevent a reoccurrence.

Additional Resources

US Chemical Safety and Hazard Investigation Safety Board (CSB), Safety Bulletin: Sodium Hydrosulfide – Preventing Harm, https://www.csb.gov/assets/1/20/sodiumhydrosulfidebulletin.pdf?13884

US Chemical Safety and Hazard Investigation Safety Board (CSB), MGPI Processing, Inc. Toxic Chemical Release, https://www.csb.gov/mgpi-processing-inc-toxic-chemical-release-/

US Chemical Safety and Hazard Investigation Safety Board (CSB), Investigation Report; Hydrogen Sulfide Poisoning (2 dead, 8 injured), https://www.csb.gov/georgia-pacific-corp-hydrogen-sulfide-poisoning/

US National Transportation Safety Board, Hazardous Materials Accident Brief – Whitehall Leather Company, Whitehall, Michigan, https://rib.msb.se/filer/pdf/17479.pdf

EHS Daily Advisor, How Chemical Deliveries Can Go Wrong, https://ehsdailyadvisor.blr.com/2018/04/chemical-deliveries-can-go-wrong/

ResearchGate, A Human Factors Review of Road Tanker Offloading Across Multiple Organizations: Simple, Complicated and Complex problems, https://www.researchgate.net/publication/337243390_A_Human_Factors_Review_of_Road_Tanker_Offloading_Across_Multiple_Organizations_Simple_Complicated_and_Complex_problems

Chemical Receiving Safety, Joel Anderson, Chief Operator, Sebago Lake Water Treatment Facility, Portland Water District, https://mwua.org/wp-content/uploads/2010/09/ChemicalReceiving.pdf

The Chemical Engineer, New safety video from CSB highlights tank fill dangers, https://www.thechemicalengineer.com/news/new-safety-video-from-csb-highlights-tank-fill-dangers/

A Safety Bulletin by Hazardex, Preventing chemical deliveries to the wrong tank, https://www.hazardexonthenet.net/article/192100/Preventing-chemical-deliveries-to-the-wrong-tank.aspx

Sofis valve operation, Ensuring safe valve line ups during chemical offloading in the water industry, https://www.sofisglobal.com/applications/ensuring-safe-chemical-offloading/

icheme, Paul Lucas, Principal Consultant, ABB Consulting, Belasis Hall Technology Park, Billingham TS23 4EB, An investigation into chemical tanker offloading and loading processes: Do they protect against the full range of predicted human error types?

The Mecklenburg Sun / SoVaNow.com, "Misplaced chemical kicks up toxic fumes," Roanoke River Service Authority water treatment plant near Bracey VA. https://www.sovanow.com/articles/misplaced_chemical_kicks_up_toxic_fumes/

October 1996, Avonmouth, UK, Albright and Wilson, Storage tank and road tanker fire due to inadvertently mixing chemicals during off-loading. https://www.hse.gov.uk/comah/sragtech/casealbright96.htm

Incident 4: Two Brothers and Two Others Were Killed by Hydrogen Sulfide While Attempting to Repair a Landfill Leachate Pump (4 Fatalities)

Superior, Wisconsin
Kimmes Demolition Landfill
Incident date: November 1, 2007

This report is assembled from the OSHA report and local media reports.

This tragic incident occurred on November 1, 2007, at around 5:30 p.m., when four employees were in the process of replacing a sump pump in an underground lift station at a local construction debris landfill. The lift station was designed to collect drainage and leached water from the landfill and process it at a local wastewater facility.

Two employees were brothers; all four were well-known and prominent community members. The brothers owned the demolition landfill and several other businesses in the local area. The sequence of events is unclear; however, what is known is that one of the brothers entered the underground pump station and commenced work on a nonfunctioning pump designed to pump liquids that accumulated (leachate). At this time, he was overcome by hydrogen sulfide gas. The second brother realized that he was in trouble and reacted by quickly entering the lift station to help, and then he also collapsed. Now, two brothers are in serious distress in the deep well, and a third person enters the confined space to support and assist the first two. As you may have suspected, he was also quickly overcome and collapsed. Now, two brothers and a rescuer are in the well, and all appear to be unconscious. The fourth person quickly descended into the well to help his other teammates but quickly collapsed. Finally, someone on the surface calls 911, requesting assistance from trained emergency responders. When the responders arrived, they found all four, two brothers and two others who were also working on the pump in the well, and, unfortunately, all four were deceased.

The responders performed air quality monitoring with an electronic sensor. They recorded hydrogen sulfide readings in the pit far above levels required to sustain human life. All four had died due to exposure to the hydrogen sulfide. At this point, the retrievals of the four bodies were initiated.

The lift station pump was described as in a pit about 20 feet deep, with a three-foot-square cover at the top. Reportedly, one of the men had entered the pit using a ladder to work on the leachate water pump at the bottom of the pit. When he collapsed, the remaining three entered one at a time to aid their fallen workmates. This is a tragic outcome; however, many other stories are similar to this one in the OSHA hydrogen sulfide incident database. Please see Attachment 2 for a copy of this database, or go to OSHA.gov and search for hydrogen sulfide incidents.

As defined in the OHSA regulation for a confined space, the lift station was clearly a permit-required confined space. Unfortunately, the company had not evaluated the lift station to determine that it was a permit-required space and had not developed and implemented a written permit space program. There was no evidence of air quality testing or monitoring, no personal protective equipment (PPE) such as self-contained breathing equipment, and no indication of a pre-planned rescue plan.

This tragic incident claimed the lives of two brothers and two others, all of whom were well-respected businessmen from the Superior area. The Kimmes brothers owned businesses in south Superior, including a construction company, an asphalt business, and a demolition landfill. They were considered pillars of the community and significant contributors to the local economy.

The sketch in Figure 7.15, from the US Centers for Disease Control and Prevention investigation report, illustrates how landfill liquids and gases, including hydrogen sulfide, can migrate from the landfill into adjacent tunnels or

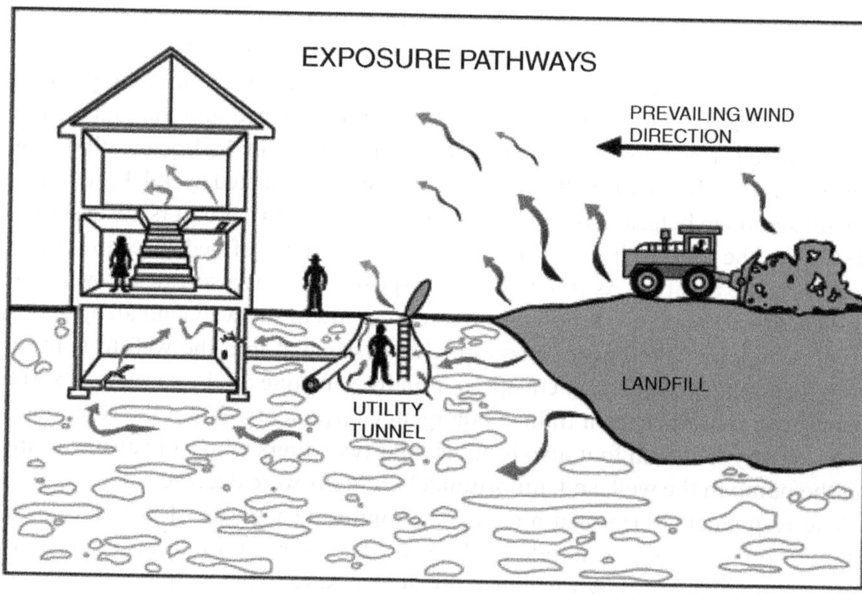

Figure 7.15 A series on Landfill Gas Safety and Health Issues from Chapter 3. The illustration indicates how landfill gases, including hydrogen sulfide, can migrate from the landfill into adjacent tunnels or pits where process equipment such as pumps may be installed. This is what happened at the J. Kimmes construction landfill site, where four fatalities occurred on November 1, 2007. See the Additional Resources section for a link to the Centers for Disease Control and Prevention publication.
Source: The US Centers for Disease Control and Prevention. Public Domain.

pits where pumps may be located. This happened in this incident, with sour water, containing hydrogen sulfide flowing into the pump pit.

Also, there was no evidence to suggest that they had been trained in confined space entry procedures, and they apparently did not recognize the hazards associated with this kind of activity. When the first brother succumbed to the hazardous environment, the three did what their instincts called for them to do. They entered the confined space to help their brothers and teammates.

This is a true tragedy, and if confined space procedures and practices had been developed, closely followed, and enforced, this should not have happened. There are other very similar cases of exactly this detailed in Appendix 2, the analysis of OSHA hydrogen sulfide reported incidents. These cases also detail would-be rescuers attempting to rescue someone who has been overcome by hydrogen sulfide and is also killed in the process. As outlined several times in this book, this should never happen. Training should include the specific actions to take in the event a teammate collapses in areas where there is a possibility of hydrogen sulfide being present. They should first stop and recognize what may have happened, then immediately call for assistance from trained professionals with the appropriate training, rescue response gear, and protective equipment, such as SCBA.

Incident 4: Key Lessons Learned

November 1, 2007: Kimmes Demolition Landfill, Superior, Wisconsin (4 Fatalities, Including 2 Brothers)

This tragic accident claimed the lives of two brothers (44 and 40 years) and two others, a Lakehead company employee (47 years) and a private electrical contractor (41 years). This incident occurred while the four men were repairing an underground pump used to pump the leachate water discharged from a construction landfill site. Unfortunately, we don't have all the information on this incident since there were no witnesses.

What we do know is that these guys were very successful businessmen in the city of Superior and the surrounding areas. They, with their father, had started and ran several very successful businesses, including this construction landfill. We also know that they were attempting to do maintenance on an underground leachate water pump that pumped the foul water that drained from the landfill so that it could be pumped to a wastewater treatment plant.

We also know that one of the men had entered the pit where the pump was located and quickly succumbed to the toxic gas in the pit and collapsed. Following his collapse, one after the other entered the pit to try and provide assistance and possibly to attempt rescue, and each one was overcome quickly, collapsed, and died due to the exposure. Tragically, all four men were killed in this incident.

When the emergency responders arrived, they conducted a gas test of the pit and found it to have a lethal concentration of hydrogen sulfide present.

This pit certainly met all the requirements of an OSHA permit-required confined space and should have been gas-tested prior to any attempted entry. The men should have had a dedicated entry attendant who could summon for rescue if needed and a documented rescue plan with rescue personnel and equipment ready to respond. Unfortunately, they had none of this and more than likely were not aware of the hazards of hydrogen sulfide. We know in hindsight that the leachate water contained hydrogen sulfide from the decaying debris and garbage from the landfill.

What I get out of this is that anytime we hear of anyone planning to enter a pit that is more than 4 feet deep, a storage tank, a tank truck, a grain silo, a manure pit, or any similar enclosure, we should discuss this with them and ensure they are aware of the hazards associated with work of this type. A gas test should be completed before entry, and all the other requirements of the OSHA permit-required confined space entry must be in place and followed. Interventions of this type can save lives. If we know that a task of this type is planned, we must be prepared to intervene when required. This can save lives.

Incident 4: November 1, 2007: Kimmes Demolition Landfill, Superior, Wisconsin (4 Fatalities, Including 2 Brothers)

Author's Comments and Perspective

This incident highlights how attempted rescues resulted in three additional fatalities in this incident. When the first person collapsed due to being overcome by hydrogen sulfide, three additional people went into the excavation, one after the other, and all three of these also succumbed to the effects of the H_2S, resulting in four fatalities in this incident.

As I stated earlier, this is a natural human urge we all have. When we see our teammate collapse, even though we may be aware that the collapse was due to hydrogen sulfide, our instinct is to jump in and try to help rescue the person who is down. This does two things. First, it delays the responder's access to the person who collapsed first. The body of the person(s) who attempted rescue is usually on top of the person who is already down, delaying the response. The second thing is that the responder is also usually killed by the effects of the hydrogen sulfide. This incident resulted in four fatalities when there should have only been one. There are case studies of even more fatalities, in which one person goes down and is quickly followed by others, usually one by one attempting rescue, who were also killed.

Appendix 2, data from the OSHA injury database, contains a listing of H_2S incidents from January 1993 to December 2023, or just over 30 years, and contains 179 deaths, 124 injuries, and 101 hospitalizations, all due to exposure to hydrogen sulfide. This is an average of six fatalities per year due only to those killed by hydrogen sulfide, and as I already said a couple of times, this is still happening today. This is the only data from the United States, and although the data for worldwide fatalities is not readily available, I don't believe it is any better.

What is interesting about this data is that it highlights 49 attempted rescues of someone who has been overcome by the effects of hydrogen sulfide. These 49 attempted rescues resulted in 71 deaths, including both the rescuers and those who were being rescued, and 45 hospitalizations, also including the rescuers. This data means that we should be very careful before attempting to rescue someone who may have succumbed to H_2S and only respond when we are properly trained and have the appropriate PPE and equipment before attempting a rescue.

The US Bureau of Labor Statistics reports that historically, there have been 46 fatalities over seven years in the United States, all due to exposure to hydrogen sulfide. This relates to about six to seven fatalities per year in the United States. They reported that this makes exposure to hydrogen sulfide gas one of the leading causes of gas inhalation deaths in the United States each year. These numbers are in line with what I have reported here.

Recognizing that this trend is continuing, this number of deaths due to hydrogen sulfide should be unacceptable.

Additional Resources

MPRNews, Toxic gas kills landfill workers (Kimmes Construction), https://www.mprnews.org/story/2007/11/02/workerdeaths

Superior, Wisconsin Landfill – Two Brothers and Two Others Killed by Hydrogen Sulfide, Incident date: November 1, 2007, Note: This incident was reported by the US OSHA in two separate reports as outlined below. The first one reports three deaths, and the fourth is detailed in the second report. All four were at the same location and occurred, essentially one after the other by entering as excavation where the atmosphere was contaminated with hydrogen sulfide gas. https://www.osha.gov/ords/imis/accidentsearch.accident_detail?id=200340743, https://www.osha.gov/ords/imis/accidentsearch.accident_detail?id=200340750

US Occupational Safety and Health Administration (OSHA), Permit Required Confined Space Regulation, 29CFR 1910.146 (Specifically 1910.146b), https://www.osha.gov/laws-regs/regulations/standardnumber/1910/1910.146#:~:text=Permit%2Drequired%20confined%20space%20)permit,period%20when%20entry%20is%20authorized.

US Occupational Safety and Health Administration (OSHA), Permit-Required Confined Spaces, https://www.osha.gov/sites/default/files/publications/osha3138.pdf

Centers for Disease Control and Prevention, Chapter 3: Landfill Gas Safety and Health Issues, https://www.atsdr.cdc.gov/hac/landfill/html/ch3.html

Incident 5: Hydrogen Sulfide Release, Valero Refinery, St. Charles, LA. The Isolation Blind Was Improperly Installed, Resulting in Release (1 Fatality, 1 Injury)

Valero Refinery – St Charles, Louisiana
Incident date: March 6, 2011 (1 fatality, 1 injury)

This report is assembled based on reports by the Louisiana State Police and the OSHA.

At this refinery, the Cat Cracker unit had been shut down for maintenance and modifications, and maintenance work was underway. At approximately 3:00 p.m. on March 6, 2011, 10 workers were performing maintenance on a 180-ft process tower. One worker and a coworker were doing site preparation, including grinding, to install a 20-inch nozzle on the tower at the 100-ft level. Five of the 10 workers were at the very top of the tower, preparing to do confined space entry for internal tower work. Three workers were at the ground level doing structural work.

The hydrogen sulfide release was detected at around 3:00 p.m. Reportedly, at least one of the contractors smelled the distinctive rotten egg odor, and the personal hydrogen sulfide (H_2S) meters on the two workers started to alarm simultaneously. They both decided to evacuate the unit to escape the release quickly and were using the caged ladders to escape. One of the workers called out on his radio, alerting the other workers, including a ground-level supervisor, of the encounter with the H_2S. Immediately, all workers on the tower began to descend from their respective levels. The first two workers began descending from the 100-ft level and encountered an increasingly dangerous atmosphere. As those workers at the very top began to descend, their personal meters also began alarming.

The ground-level superintendent suspected that the H_2S was at a higher concentration near the 100-ft level and called out on the radio for those workers who were descending from the very top to go back up. The first two workers began to lose consciousness and fell from their respective locations. One worker fell to the deck near the 100-ft level. The other worker slid down the ladder cage that connected the deck near the 100-ft level to the deck below. When he awoke, he was hung on the caged ladder by his clothing, and he quickly descended to safety. He slid down into a more breathable atmosphere, sustaining bruising, but survived with minor injuries. However, the other employee passed out from exposure to the H_2S, fell, and suffered blunt force trauma to his head from the fall.

Both injured workers were transported to a local hospital at 3:35 p.m. At approximately 6:04 p.m., Valero made notification that two contract workers had been exposed to hydrogen sulfide and, as a result, had sustained injuries from falls. Both contractors were sent to the hospital for treatment. However, one was pronounced dead shortly after arrival due to injuries sustained in the fall.

According to reports, the Louisiana State Police met with Valero representatives the same day, and representatives from the OSHA began an investigation into the incident. According to reports, it appears that isolation blinds had been improperly installed, resulting in the release of the hydrogen sulfide during the steaming of the piping system. The release occurred in the area where the contractors were working, apparently resulting in the contractors being exposed to the hydrogen sulfide.

Reportedly, it took approximately 93 hours for the company employees to locate the source of the release and secure the leaking pipe. It was also reported that no significant amount of H_2S was detected during this period. It was also determined that no significant quantity of H_2S was released to the surrounding environment and that the public had not been affected. However, all ongoing work in the vicinity ceased in order to allow the investigation and to identify the source of the leaking toxic gas.

After the release source was identified, the improperly installed blind was replaced with observation by the Louisiana State Police. Monitoring was also done to confirm that the new blind was not leaking. The investigation also planned to include recommendations to prevent a recurrence of this event or a similar one. This incident resulted in a loss of containment of hydrogen sulfide, an extremely toxic gas, one worker fatality, and one worker injury. The injured worker was treated and released from the hospital after treatment.

Although a turnaround is fast-moving and can be a confusing atmosphere due to the fast pace and the number of simultaneous activities, every blind must be tracked on a blind list or another tracking document. Every blind that is installed and every blind that is removed must be accounted for with the tracking tool. This incident also exemplifies that before a work permit is issued to open any flange to install a blind, the location must be identified with a tag or otherwise identified as the correct blind location. A walk-down must also be done to ensure that blind installation is correct and the vessel is at a zero-energy state before any flanges are opened or manways are removed.

This incident also highlights the importance of full compliance with the OSHA standard for energy isolation and lockout–tagout provisions of the standard. This equipment does not appear to have followed the standard since the system was not depressurized or properly isolated and continued leaking hydrogen sulfide-containing gas into the environment where personnel were working. Again, this highlights the importance of a walk-down to ensure every isolation point is properly isolated and secured with a blind or another device to prevent the loss of containment of flammable or toxic material.

I also cover this in more detail in the textbook "*Process Operations Safety: The What, Why, and How Behind Safe Petrochemical Plant Operations.*"

Incident 5: Key Lessons Learned

March 6, 2011: Valero Refinery, St Charles, LA (1 Fatality, 1 Injury)

The fluidized catalytic unit at this refinery was undergoing shutdown and maintenance when 10 workers were working in the structure on a 180-foot-tall fractionation column when hydrogen sulfide was released from an open pipe in the structure. The release occurred somewhere above the 100-foot level, where a blind flange had been improperly installed, leaving the pipe open. Two of the workers were exposed to hydrogen sulfide; both collapsed and fell due to H_2S exposure. They were sent to the hospital for treatment, where one died due to blunt force injuries he sustained in the fall.

The investigation found that the release occurred due to an improperly installed blind flange, which left the process line in the open position. The H_2S release occurred when operations began injecting steam into other piping in the structure to clear the piping for maintenance work.

As stated earlier, this incident highlights the importance of closely following the OSHA standard for equipment isolation and lockout/tagout (29 CFR 1910.147), "The control of hazardous energy (lockout/tagout)." Specifically, 1910.147(a)(3)(i) requires employers to "develop a program and to utilize procedures for affixing appropriate lockout devices or tagout devices to energy isolation devices, and to otherwise disable machines or equipment to prevent unexpected energization, start-up or release of stored energy in order to prevent injury to employees." Also, 1910.147(c)(4)(i), "Procedures shall be developed, documented and utilized for the control of potentially hazardous energy when employees are engaged in the activities covered by this section."

This means that a detailed procedure must be in place for equipment isolation, including blinding. A walk-down must be done to ensure that blinds are properly installed at the correct location and that equipment is properly isolated to guard against "stored energy."

This walk-down is extremely important, and it must be done between the operations group and the people who are doing the work. Blind points must be identified on the unit drawings and at each location in the field (on the unit). The operations group should tag blinding locations to indicate that the piping has been prepared and the flange is ready for opening (for blind installation or removal). As indicated, this should be a weatherproof and tamperproof tag used only for this purpose.

Additional Resources

Louisiana Refinery Accident Database, Valero (26003), Norco, LDEQ Accident Report, Accident # 129748, https://louisianarefineryaccidentdatabase.org/incident.php?serno=4736

US Occupational Safety and Health Administration (OSHA), Accident Report Detail, Accident Summary Nr: 200556389, Employee Is Overcome by Hydrogen Sulfide and Dies in Fall, https://www.osha.gov/ords/imis/accidentsearch.accident_detail?id=200556389

Worker death at Valero Refinery in Norco highlights ongoing problems, Posted on March 11, 2011 by Louisiana Bucket Brigade, https://labucketbrigade.wordpress.com/2011/03/11/worker-death-at-valero-refinery-in-norco-highlights-ongoing-problems/

US Occupational Safety and Health Administration (OSHA), 1910.147 – The control of hazardous energy (lockout/tagout). https://www.osha.gov/laws-regs/regulations/standardnumber/1910/1910.147

Incident 6: Three Utility Workers Killed by Hydrogen Sulfide While Working in Municipal Sewer System (3 Fatalities, 5 Injuries)

Long Key Road (Lake Surprise Estates Subdivision)
Key Largo, Florida
Incident Date: January 16, 2017

This report was assembled from the OSHA incident report and several local media reports of this tragic incident.

Nearby residents reported odors that smelled like rotten eggs had been in the area for several months and reported that there had been some issues with poor sewer drainage in the Lake Surprise Estates subdivision. Three local utility workers were investigating the area, and one noticed the pavement was loose. While checking an underground pipeline, 34-year-old Elway Gray opened an adjacent manhole in a newly paved section of the road and climbed in to investigate further. Reportedly, almost as soon as he entered the manhole, he collapsed and became unconscious. Quickly, a 49-year-old laborer, Louis O'Keefe, followed Mr. Gray into the manhole to provide aid and assistance, and he also collapsed. With the two workers in the manhole and both unconscious, a 24-year-old equipment operator also entered the manhole to help his coworkers.

Unfortunately, all three of the utility workers perished because of exposure to the toxic hydrogen sulfide. Two other workers and a local firefighter were also exposed. Leonardo Moreno, who has been with the Key Largo fire department for eight years, reportedly made a desperate effort to retrieve the men, but unfortunately, he also succumbed to the effects of the hydrogen sulfide in seconds and passed out. He was flown to a local hospital, where he was reported to be in critical condition. Gas testing of the environment inside the manhole revealed lethal levels of hydrogen sulfide (H_2S) and carbon monoxide (CO).

Three sheriff's deputies were also taken to a local hospital for observation and treatment after inhaling the toxic vapors. Several homes were evacuated, and the Miami-Dade County Hazmat Team conducted atmospheric monitoring in the area, which confirmed the presence of methane gas and hydrogen sulfide.

The manway was reported to be just wide enough to fit a body. It had filled with H_2S and methane from decaying vegetation, according to the *Miami Herald* newspaper. None of the four men, the three who died and the firefighter who attempted a rescue, were wearing an SCBA. It is believed if they had, their lives would have been saved. Monroe County Sheriff Rick Ramsey was quoted as saying that the firefighter, Leonardo Moreno, could not fit in the manway while wearing his SCBA. After he entered and collapsed, another firefighter was able to enter the manhole wearing his SCBA and rescued Mr. Moreno. The firefighters also recovered the bodies of two of the workers. The third fatality was recovered the following afternoon.

The OSHA investigated the incident and found that safety failures led to the deaths. The OSHA citation noted several violations, including:

- Lack of atmospheric testing (which would have detected the H_2S)
- No confined space entry program (which would have required atmospheric testing)
- No confined space entry permits by a qualified person (which would have required training for all entrants)
- No rescue plan
- Rescue services not available in a timely manner
- Rescue equipment not available at the site

The US Department of Labor announced in a news release Tuesday that Douglas N. Higgins Inc. and its related contracting company, McKenna Contracting LLC, were being cited for 10 serious violations in connection with the January 16 deaths of three workers. The imposed OSHA citations were for "failing to purge or ventilate the confined space before entry, exposing the workers to an asphyxiation hazard, and not providing necessary rescue and emergency equipment."

The OSHA Regional Director, Condell Eastmond, was quoted as saying, "Three employees needlessly lost their lives and others were injured due to their employer's failure to follow safe work practices."

Incident 6: Key Lessons Learned

Three Utility Workers Killed by Hydrogen Sulfide While Working in Municipal Sewer System

Long Key Road (Lake Surprise Estates Subdivision)
Key Largo, Florida
Incident Date: January 16, 2017

- Always, Always, Always, follow the OSHA Confined Space Entry regulation when entering a process vessel, an excavation, a pump pit, or other underground area or enclosure not designed for human access.
- This means a test of the atmosphere for O_2 (oxygen), H_2S (hydrogen sulfide), LEL (flammability or explosivity), and any other toxic specific to the process.
- Ensure that a rescue plan has been developed and resources are readily available to execute the plan if needed.
- Entrants should always wear a harness and lifeline unless this makes the task unsafe. If unsafe, a risk analysis should be completed, and an alternate plan or approach developed.

This is another tragic case of workers entering what should have been identified as an OSHA-defined permit-required confined space without following any of the procedures or requirements outlined in the OSHA standard. I'm sure they did not think about the confined space standard or identify hydrogen sulfide as a significant hazard before opening the excavation and entering. One worker entered the excavation and reportedly succumbed within seconds. He was quickly followed by a second worker who obviously experienced what I describe as the human urge to rescue and was also overcome. Then, the third worker also entered to help the first two, and he also collapsed in seconds due to the very high concentrations of hydrogen sulfide.

All three of these workers were killed by the toxic hydrogen sulfide that had accumulated in the confined space. Two other workers and a local firefighter were also exposed, including one who was very seriously injured. Three sheriff's deputies were also taken to a local hospital for observation and treatment after inhaling the toxic vapors.

When emergency responders arrived, gas tests revealed that lethal levels of hydrogen sulfide (H_2S) and carbon monoxide (CO) had accumulated due to decaying vegetation that was trapped in the manhole.

Additional Resources

US Occupational Safety and Health Administration (OSHA), OSHA News Release – Atlanta Region, https://www.osha.gov/news/newsreleases/region4/07182017

US Occupational Safety and Health Administration (OSHA), Confined Space Entry Standard, 29 CFR 1910.146 – Permit-required confined spaces

Miami Herald News, Three men descend to their death, overcome by poisonous gas under the ground, https://www.miamiherald.com/news/local/community/florida-keys/article126799319.html

FLKEYSNEWS, Report: Series of events led to utility workers' deaths, https://www.flkeysnews.com/news/local/article151487457.html

Incident 7: ConocoPhillips Company: Employee Found Unresponsive While Attempting to Drain Liquids from a Tank in a 90 Sq ft Oil Well Shack (1 Fatality)

RTB Well Site
Watford City, ND
Incident date: February 24, 2023 (1 Fatality)

This incident description, found in Appendix 2, is developed from the OSHA H_2S incident database with supplemental information from local media reports.

At approximately 12:00 p.m. on February 24, 2023, a ConocoPhillips employee performed well-site pumping operations at a company-operated oil and gas facility near Watford City, ND. One of the employee's coworkers had made several unsuccessful attempts to contact the employee to check on him.

Upon the coworker's arrival at the well site, he found the employee unresponsive inside a small (approximately 9 feet by 10 feet) oil well shack. The employee had been working inside the shack with the entry door closed while attempting to drain liquids from a tank designed to separate water from hydrocarbons (produced water and condensate). The gases are also separated and are sent to the flare. It was apparent to the coworker that the employee was attempting to unplug a bleeder on the pump when he found a hose connection made to a bleeder on the pump discharge and an overturned bucket on the floor. The overturned bucket had a small amount of liquid and a puddle of liquid on the floor from the end of the hose. It was known that these gases and fluids may be rich in hydrogen sulfide.

The coworker responded by pulling the unconscious employee from the building and quickly called for an emergency response. He also conducted CPR while he waited for medical help to arrive. The coworker reported that the employee had foam in his mouth, and blood was coming out of the employee's mouth while he performed CPR. The coworker also observed that the employee had what appeared to be chemical-type burns. When the responders arrived, they transported the employee to the hospital, where he was pronounced dead. The employee was overcome by highly flammable gases, including potential hydrogen sulfide overexposure.

Incident 7: Key Lessons Learned

February 24, 2023: ConocoPhillips Company, RTB Well Site, Watford City, ND (1 Fatality)

I thought about this incident before deciding to include it here. However, since this shows just how quickly these incidents can occur and how tragic they can be, I felt it would be appropriate to include it. One slip in judgment, or not following established procedures, is all that is required for one, and sometimes several people to be killed. I don't have additional details on this incident and don't have access to company policies or safety requirements. However, from my perspective, the following should always be required for this kind of activity and must always be enforced by management. This will sound awfully like other incidents you have already reviewed in this book. Some are repeated here intentionally because they are that important.

- Always wear a personal H_2S monitor, and if it goes into alarm, immediately leave the area and return only if you are wearing proper PPE, including SCBA.
- Although this process shack was not designated as a confined space and may not meet the definition of a confined space, as soon as the employee started draining product from the bleeder, he should have ensured adequate ventilation.

  Working with the door closed turned this into the equivalent of a confined space with little, if any, ventilation. He also should have been prepared to encounter H_2S with proper PPE and procedures. For H_2S atmospheres, the PPE should always include an H_2S monitor and an SCBA.
- Follow the OSHA Confined Space Entry regulation when entering a process vessel, an excavation, a pump pit, or other underground area or enclosure not designed for human access or occupancy.
- This means a test of the atmosphere for O_2 (oxygen), H_2S (hydrogen sulfide), LEL (flammability or explosivity), and any other toxic specific to the process.
- When entering a confined space, ensure that a rescue plan has been developed and that resources are readily available to execute it if needed. Entrants should always be wearing a harness and lifeline unless this makes the task unsafe. If it is deemed unsafe, a risk analysis should be completed, and an alternate plan or approach should be developed.
- It is always important to ensure proper ventilation when working in a confined space or a process shack, like what was being done in this case, anytime flammables or hydrogen sulfide may be present.

Additional Resources

US Occupational Safety and Health Administration, Inspection: 1653851.015 – ConocoPhillips Company, Inspection Detail | Occupational Safety and Health Administration osha.gov

US Occupational Safety and Health Administration (OSHA), Confined Space Entry Standard (29 CFR 1910.146), https://www.osha.gov/laws-regs/regulations/standardnumber/1910/1910.146

Incident 8: Worker Exposed to Hydrogen Sulfide and Falls from the Top of the Truck (1 Fatality)

Randy Berdahl Trucking, Ltd (Eckville, Alberta)
Incident location: Medicine River Oil Recyclers, Near Eckville, Alberta (Canada)
Incident date: November 2, 1999 (1 Fatality)

Alberta Workplace Health and Safety investigated this incident and filed a report. This is a summary of the report written as interpreted by the author and additional information from local media reports. Please see the link in the Additional Resources section for a copy of the report. Workplace Health and Safety also issued an order that required the trucking company, the oil recycling company, and Samson Canada Ltd., the company that owns the production wells, to investigate the incident and implement corrective measures to prevent a reoccurrence. Workplace Health and Safety indicated in their report that the companies complied with the order. However, it is unclear what steps were taken to prevent a recurrence of this incident.

A summary of the Berdahl Trucking incident that resulted in the fatality:

This incident occurred at the Medine River Oil Recycling company near Eckville in Alberta Province in Canada on November 2, 1999. It resulted in the fatality of a truck driver who was delivering a product to be recycled.

Samson Canada Ltd. operates oil and gas production facilities in Alberta, and their process typically involves reinjecting produced sour water from the wells back into the formation. However, a failure occurred on one of their produced water injection pumps at the Condor Battery site. As a temporary measure, Samson Canada Ltd. contracted Randy Berdahl Trucking Ltd. to transport four loads of produced sour emulsion from their produced sour water storage tank to the Medicine River Oil Recyclers.

At about 4:30 p.m. on November 2, 1999, a truck driver, also the trucking company's owner, delivered a load of produced sour water from an Alberta oil production facility to the oil recycling plant. This product was an emulsion consisting of some hydrocarbons and sour water (water containing hydrogen sulfide and other sulfur compounds) produced from Alberta's oil and gas wells.

The truck configuration was a truck connected to a large tank trailer coupled with a smaller tank trailer referred to as a pup tank. There was a line of trucks ahead of him waiting to unload; therefore, the driver climbed to the top of the large tank and collected a sample of the large tanker using an extraction tool designed to collect samples of produced sour water and emulsion from the tank truck and pup tank.

After collecting a sample from the truck tank, he proceeded to the top of the pup tank to collect another sample. When the truck driver opened the hatch on the pup tank, he was exposed to hydrocarbon vapors and hydrogen sulfide, which was released from the pup tank. After exposure, the truck driver collapsed and fell approximately 10 feet (3 meters) to the ground and sustained multiple blunt force trauma injuries, resulting in his death.

A witness saw the driver fall from the truck and provided first aid. We don't know if this included attempted resuscitation in any form. Emergency medical services were called and arrived at 4:45 p.m., and the driver was quickly transported to the regional hospital. The driver was then airlifted to a Calgary hospital, where he was pronounced dead at 1:40 a.m. the following morning.

Samples taken from the truck tanks at 10:00 p.m. on November 2, 1999 (6.5 hours after the incident) revealed H_2S concentration of 326 ppm in the tank truck and 279 ppm in the pup tank. The production facilities where the trucks were loaded are located about 6 miles (10 km) South of Condor, Alberta, where produced oil/gas/water emulsions are collected from seven well sites. The produced fluids from this formation are known to have H_2S concentrations of 7% H_2S. Seven percent H_2S is equivalent to a concentration of about 70,000 ppm, or enough that one breath results in instantaneous death. A concentration of 279 ppm should not result in immediate knockdown; however, we don't know what the H_2S concentration was in the vapor space at the time of the incident.

The Alberta Workplace Health and Safety Office concluded that the primary cause was that the driver was "exposed to hydrocarbon vapors and hydrogen sulfide and fell approximately 3 meters (9.8 feet) to the ground." They concluded that the secondary cause was "the truck driver was not protected from hydrocarbon vapors and hydrogen sulfide with respiratory protective equipment."

Actions taken in response to this workplace fatality (as reported by Alberta Workplace Health and Safety)

Randy Berdahl Trucking voluntarily stopped transporting produced sour emulsions containing hydrocarbon vapors and hydrogen sulfide. Workplace Health and Safety reported that "the company complied with the order issued by Workplace Health and Safety."

Samson Canada Ltd. implemented a code of practice for transporting oilfield fluids containing hydrogen sulfide. Workplace Health and Safety reported that they complied with the compliance order that had been issued. It is unclear what additional precautions or changes in requirements for personnel H_2S safety monitoring and/or PPE were made in the new code of practice.

Medicine River Oil Recyclers conducted an incident investigation and prepared the report. The company also developed and implemented a code of practice for cargoes containing hydrogen sulfide. Workplace Health and Safety reported that the company complied with the compliance order that had been issued. It is unclear what additional precautions or changes in requirements for personnel H_2S safety monitoring and/or PPE were made in the new code of practice.

Workplace Health and Safety reported that their organization responded to the scene and commenced an investigation on November 2, 1999. They issued an order requiring the trucking company and the oil recycling company to conduct an incident investigation. They ordered both companies to implement corrective measures to prevent reoccurrence. They also reported that they completed an inspection at the oil production facility to ensure workers were adequately protected against exposure to hydrocarbon vapors and hydrogen sulfide.

Although Workplace Health and Safety ordered both companies to implement corrective measures to prevent recurrence, it is unclear what changes have been made or will be made to prevent a recurrence of this incident.

Additional considerations for this event (from the author's perspective)

- Anytime personnel are working around process vessels and equipment containing H_2S at high concentrations (this means concentrations where the workers could be exposed to concentrations above 10 ppm) (the NIOSH PEL), they should be wearing an H_2S personal monitor. When the alarm sounds, they should be trained to immediately leave the scene and return to troubleshoot only after donning an SCBA.
- For each task involving vessels or piping containing H_2S at these concentrations, every employee and contractor should be trained on the hazards of H_2S, and training should also provide how to do a "Last Minute Risk Assessment" (LMRA) to:
  1. Identify the potential hazards, including other job site hazards and H_2S.
  2. Develop a plan to mitigate the identified hazards and implement the plan. This includes H_2S, fall protection, and rescue.
  3. Consider the PPE required to conduct the task, especially for H_2S.
  4. Ensure they wear the correct PPE BEFORE starting the task.
- With the potential for H_2S outlined in this incident, truck drivers should be wearing either hose-line supplied breathing air (breathing air that has been certified as Grade D breathing air) or SCBA before they open a hatch on a tank truck or any process vessel such as a storage tank or drum.
- In addition to the above, companies should also provide driver training for body position when opening hatches. Avoid leaning over an open tank hatch and stand away or upwind when possible. See the link to a safety article by *EHS Daily Advisor*, "What You Can Do to Avoid Vapor Hazards When Opening a Tank Hatch," for additional considerations.

- Companies and drivers should also provide fall protection for personnel when working on the top of a tank truck (without guard rails or other fall prevention methods). This is required by US OSHA regulation: OSHA Regulation 29 CFR 1910.23(c) – (General Industry) says, "Every open-sided floor or platform 4 feet above an adjacent floor or ground level shall be guarded by a standard railing on all open sides except where there is entrance to a ramp, stairway or fixed ladder." While OSHA requires this, it is good practice for companies outside the United States.

Incident 8: Key Lessons Learned

November 2, 1999: Randy Berdahl Trucking Ltd. (Medicine River Oil Recyclers), Eckville, Alberta (Canada)

As I said earlier, I started not to include this incident. However, it clearly illustrates just how quickly an incident like this one can happen and how tragic it can be.

The truck driver's task was simple: climb to the top of the truck's tank trailer and collect a sample of the product he was delivering to the waste recycler. Simple – until you stop and think about the potential hazards and how those hazards can be controlled. I refer to this as an "LMRA," where the people doing the task stop and think through the potential hazards and how those hazards should be addressed.

Since this is so important, let's talk briefly about LMRA and how it works:

- Why should we do an LMRA?
  An LMRA helps identify, eliminate, or reduce risks that may result in an accident, injury, or environmental incident on the job site. It also helps make workers aware of their surroundings before starting a new task and identifies risks that should be controlled before work starts.
- When should an LMRA be done?
  An LMRA should be done before starting any task that may present risks to safety, health, or the environment.
- Who should do an LMRA?
  Anyone can do an LMRA, but it is aimed at workers who are doing higher-risk tasks, such as a process operator doing process tasks, someone working in construction or building trades, anyone working with process vessels containing or that have contained hazardous materials, electricians working with electrical equipment, and others in higher-risk areas or doing higher-risk tasks.

The LMRA is also for someone who is doing a repetitive task or one that seems routine. The process can identify hazards that have been taken for granted but can result in severe injury or even death.

- What is an LMRA, and how could it be described?
  An LMRA is a brief review of the task about to be performed to identify all potential hazards that should be addressed before executing the task. The review should help identify any hazards that were previously not recognized or were unknown. This allows the identified hazards to be addressed by the workers before starting the task.
- How is the LMRA done or executed?
  Before starting a new task, the worker(s) should familiarize themselves with the relevant procedures and other documents associated with the task. These include the permit to work, existing risk assessments for this task or similar tasks, existing procedures for this task or similar tasks, and the Safety Data Sheets for any chemicals that may be involved in the task.

Had the truck driver done an LMRA, some of the hazards that should have been identified include the following (not intended to be a complete list):

1. Is the truck properly parked, and are the parking brake set and wheel chocks in place?
2. Is the truck properly electrically grounded and bonded to protect against static electricity?
3. How do I protect myself from a fall from the top of the tank truck?
4. What is the toxic nature of the product being delivered?
5. How do I protect myself from this toxic material?
6. Should I wear an H_2S detector when collecting this sample?
7. Should I wear SCBA when collecting this sample?
8. When collecting the sample, should I consider the wind direction or a certain body position?

If these hazards had been identified and addressed with additional PPE or a different procedure, the driver may be alive today.

The Key Lessons Learned in this incident, the LMRA, applies to essentially all other incidents covered in this textbook. This is a very important learning, and it is a very simple process. However, it can and has saved lives. Please consider this the next time you are about to perform a task that includes safety, health, or environmental hazards. Do not become a statistic or a topic in a future textbook like this one. By using the LMRA process, your life may be saved.

Additional Resources

Worker Exposed to Hydrogen Sulphide and Falls from the Top of the Truck, Alberta Health and Safety Report, https://ncsp.tamu.edu/reports/GA/hydrogen_sulphide.pdf

Newspapers.com, Truck Driver Killed – Overcome by Sour Gas, https://www.newspapers.com/article/red-deer-advocate-dell-randolph-berdahl/39934610/

British Health and Safety Executive (HSE), Managing risks and risk assessment at work, https://www.hse.gov.uk/simple-health-safety/risk/

National Library of Medicine, National Center for Biotechnology Information, Towards the next generation of LMRA instruments: the influence of generic and specific questions during risk assessment, https://pubmed.ncbi.nlm.nih.gov/31813344/

US Occupational Safety and Health Administration (OSHA), OSHA Fall Protection Compliance for Accessing Hatches on Top of Tank Trailers, https://www.linkedin.com/pulse/osha-fall-protection-compliance-accessing-hatches-top-billy-kutivan

S&S Technical, Tank Trucks & Tank Cars – Falls from Height, https://www.skidsolutions.com/blog/tank-trucks-tank-cars-falls-from-height/#:~:text=OSHA%201910%20Regulations&text=These%20regulations%20are%20in%20place,systems%2C%20and%20fall%20protection%20training

British Health and Safety Executive (HSE), Managing risks and risk assessment at work, https://www.hse.gov.uk/simple-health-safety/risk/steps-needed-to-manage-risk.htm

HSE LIFE NL, LAST MINUTE RISK ANALYSIS (LMRA), https://hselifenl.com/wp-content/uploads/2017/07/HSElife-NL-LMRA-A6-UK-2017.pdf

Incident 9: Worker Fatally Injured After Exposure to Hydrogen Sulfide While Attempting to Inject Methanol into an Oil and Gas System (1 Fatality)

Paramount Resources Ltd.
Incident Location: Simonette Oil Battery (located approximately 50 miles [80 km] northwest of Fox Creek in Alberta Province)
Incident date: January 27, 2018 (1 Fatality)

As I was wrapping up the discussion on the previous incident, the trucking company fatality that occurred at a recycling facility in Alberta, I accidentally ran across another H_2S fatality that also occurred in Alberta just a short few years ago. This one occurred at another oil and gas well site and was also investigated by Alberta Health and Safety, and a report was written.

When reviewing the report, I was intrigued when I did not see a series of clear recommendations or follow-up actions to address the incident findings. In fairness, there may be more to this than I see in the report. Also, in fairness, the Alberta Government did forward the report to the Alberta Justice Department; however, Alberta Justice came back with "no charges recommended." I'm left with how or what do we do to prevent this from happening again. The following Additional Resources section includes a link to the Alberta Health and Safety Final Report. There are some nice images in the report; you can access them at the link in the Additional Resources section. Since the report was classified as "public," I included a couple of the images here. Otherwise, I wrote this in my own words, referencing the Alberta Health and Safety report. I firmly believe that incidents of this type should be broadly shared to help prevent future incidents and lives lost due to exposure to hydrogen sulfide and other chemicals in our workplace while protecting private or proprietary information.

A summary of the Paramount Resources incident that resulted in the fatality:

Paramount Resources Ltd. (Paramount) is an oil and gas production company with a background in developing and producing natural gas, natural gas liquids, and crude oil in several provinces in Canada, including Alberta, where this incident occurred. The operator killed in this incident was well-experienced in oil and gas, including in handling sour products and products with high concentrations of hydrogen sulfide (H_2S).

During routine operations, the operator was discovered unresponsive by a coworker. He had been working at the Simonette Oil Battery, which is located approximately 50 miles (80 km) northwest of Fox Creek. It was obvious that he had been attempting to unplug a needle valve associated with the cyclone separator, which is designed to remove free water from where methanol was injected into the system. Methanol is typically used as a deicing or freeze-prevention chemical. The tubing had been removed from the needle valve, and sour gas was being released from an open valve into the confined area where the operator's body was discovered. The cause of death was attributed to poisoning by hydrogen sulfide gas.

The report indicates that when the coworker found the operator unresponsive and lying on the ground inside the partially open enclosed area (the utilidor), he shook the operator, trying to get him to respond. He then heard the gas escaping, retreated, and called his supervisor. At this point, the report says he and the supervisor donned SCBAs and pulled the operator to a safe area. They then removed their SCBAs and "attended to the operator." It is unclear if they attempted artificial respiration or what services were rendered.

Emergency medical services were called and attempted to resuscitate the operator as soon as they arrived. Unfortunately, the operator could not be resuscitated and was pronounced deceased.

The report doesn't mention the expected concentration of the H_2S at this site, although we know it is extremely high. In the previous incident in Alberta, the concentration at the well site was reported to be as high as 7% (70,000 ppm). Remember, the Immediately Dangerous to Life and Health (IDLH) for H_2S is only 100 ppm. One breath above about 1,000–2,000 ppm can result in instantaneous death.

Also, the report doesn't mention what H_2S monitoring or safety requirements are in place at this site or other similar oil and gas production facilities throughout the region and company. From my perspective, these workers should be wearing

Courtesy of Alberta Health and Safety Report No-F-OHS-137943-3AC68, November 2020

Figure 7.16 The Simonette Utilidor cyclone separator where the operator was found.

H_2S personal monitoring. With a gas monitor, the workers who entered the space would have known that they were entering a high H_2S area and that additional protection was required. It is hard to know without documentation, but I think the operator who was killed more than likely was not wearing an H_2S detector.

It is also evident that the operator who was killed was not wearing SCBA, even though he was opening tubing associated with a high-concentration H_2S system. He would be alive today if he wore his H_2S detector and an SCBA. Figure 7.16 shows the partially enclosed structure where the cyclone separator was located, and the location where the Operator was found after being exposed to hydrogen sulfide. What are the requirements for these facilities, and is there a policy covering them? I don't like the answer I conclude when it is not documented.

The Government of Alberta report included findings or conclusions and no recommendations for improving company policies, standards, procedures, and PPE requirements. There were also no recommendations for the companies involved to modify facilities or equipment.

The path forward was to send the report to the Alberta Justice Department, which was basically rejected. Again, I realize that the report may not be the full story here, and there may be action outside of the report that I am unaware of. My concern here is that there appears to be a lack of enforcement or other actions to prevent another incident like this one from occurring again. The Investigation Report summarizes what happened with little focus on improving or enforcing governmental safety policies or standards. Figure 7.17 shows the open needle valve used for methanol injection, located inside the cyclone separator building, which was the source of the hydrogen sulfide.

Courtesy of Alberta Health and Safety Report No-F-OHS-137943-3AC68, November 2020, Classified as "Public." Annotation by the Alberta Investigation Team

Figure 7.17 Inside the cyclone separator building. "A" indicates the open needle valve used for methanol injection and the source of the hydrogen sulfide.

Incident 9: Key Lessons Learned

January 27, 2018: Paramount Resources Ltd.

Simonette Oil Battery, Approximately 50 Miles NW of Fox Creek, Alberta, Canada

When the operator discovered the deceased employee got out of his truck, he heard the gas blowing from an open valve inside the partially enclosed area. This is a case where an experienced oil and gas worker was working inside a partially enclosed area (a utilidor) and had opened the process by removing a needle valve to unplug a methanol injection point. When his body was found, sour gas was released into the partially enclosed area where his body was located. The deceased employee had been attempting to unplug a methanol injection point used for freeze prevention when he was overcome by H_2S and killed. It is also clear that he had opened the process and released sour gas (gas containing high concentration of H_2S) into the partially confined space with little ventilation.

This is a case of not having or not wearing the proper protective equipment for working in a process area where hydrogen sulfide is present. There is no mention of an H_2S monitor or an SCBA, which should be required to do this work. There is no question that these workers should be wearing an H_2S monitor at all times. An SCBA should be worn anytime equipment containing high concentrations of H_2S is being opened.

Again, it is the company's responsibility to have policies requiring the use of this protective equipment and to enforce their use at the job site. The Government of Alberta's investigation also lacks details and fails to include recommendations to address these deficiencies in the Investigation Report.

The Alberta Investigation Report mentions that "a review for enforcement action" was completed on July 29, 2019, and it was determined that the file would be referred to Alberta Justice for review. The report says the entire file was sent to Alberta Justice on September 19, 2020. Alberta Justice returned a decision of "no charges recommended." I recognize that more details may be in this entire file, to which I don't have access.

Other incidents like this one, also in Alberta, Canada

Near Fox Creek, Alberta, Canada (January 6, 2011)

Release during pipeline maintenance (1 fatality, 1 injury)

The Red Deer Advocate reported that on January 6, 2011, the Royal Canadian Mounted Police (RCMP) reported three pipeline maintenance workers were exposed to hydrogen sulfide off Highway 46 near Fox Creek. Local firefighters with SCBA were able to rescue two of the workers; one was hospitalized, and another worker died from the exposure. An RCMP officer conducting a roadblock in the area was also exposed and was treated in the hospital. Alberta Workplace Health and Safety indicated they were investigating the incident.

Southwest of Whitecourt, Alberta, Canada

Release during well suspension – January 11, 2012 (11 fatality, 2 injuries)

This fatality occurred at a lease site registered with Sinopec while contract firms were suspending the well. The well was located about 25 miles (41 km) southwest of Whitecourt. Workers were suspending the well from production by inserting a blind ram into the BOP valve and sealing off the well. This controls the well during maintenance periods.

The plan for this work had identified the H_2S concentration in the well to be 11% or 110,000 ppm. However, workers were not using supplied breathing air systems, even though these systems were readily available. A release of H_2S occurred about the same time the blind ram was inserted into the BOP. The floor hand who was working adjacent to the blind ram quickly exited the deck and collapsed, and was later pronounced dead at the hospital. Two others were treated for H_2S exposure.

Incident 9: Additional Key Lessons Learned

Release during well suspension (January 11, 2012)

Southwest of Whitecourt, Alberta, Canada (Continued)

Author's Thoughts/Perspective

This is clearly another case where if the employees had been wearing SCBA, their lives would have been saved. They were working on a system containing H_2S in concentrations known to be instantaneously lethal. One breath at these concentrations results in instant death! From my perspective, wearing SCBA should have been a requirement, and it should have been mandatory, for this kind of task.

My point here is that each of these cases is a tragedy. We must learn from these and implement the learnings in our policies, procedures, and safety management systems. We must be aligned on the use of all our safety systems and processes, like mandatory use of H_2S detectors, mandatory use of SCBA when working on high H_2S systems, detailed incident investigations with strong recommendations to prevent recurrence, periodic mandatory safety training, and regulatory enforcement with strong actions when policies are violated, etc.

Once these policies and procedures are implemented, they must be enforced. Compliance with the most critical should be considered a condition of continued employment. I know that this sounds a bit harsh, but it will save lives. I realize that it is difficult to prove a negative, and when an incident doesn't happen, it is hard to say why it didn't happen. However, I believe these stringent policies and procedures will save lives.

We must also be open about sharing what has happened as broadly as possible. This message applies to this incident, to all others in this book, and to those yet to happen.

Additional Resources

Alberta Health and Safety, Investigation Report Worker fatally injured after exposure to Hydrogen Sulphide, January 27, 2018, https://open.alberta.ca/dataset/8f19f7ba-919e-4114-9fed-796bb6900439/resource/76746a13-a87c-4372-badd-7e70077a38f0/download/lbr-fatality-report-no-f-ohs-137943-3ac68.pdf

US Occupational Safety and Health Administration (OSHA), 29 CFR 1910.134 – Respiratory protection. https://www.osha.gov/laws-regs/regulations/standardnumber/1910/1910.134

Draeger, Hydrogen Sulfide – Knowledge can save lives, https://www.draeger.com/Content/Documents/Content/H2S%20Booklet-bk-PDF-9088-en-master-1810-1.pdf

Safeopedia, SCBA 101 – Meet the Respirator that Will Save Your Life, https://www.safeopedia.com/scba-101-an-introduction/2/6438

Red Deer Advocate, Worker dies after hydrogen sulfide leak at oil well site, https://www.reddeeradvocate.com/uncategorized/worker-dies-after-hydrogen-sulfide-leak-at-oil-well-site-7054098

EHS Daily Advisor, What You Can Do to Avoid Vapor Hazards When Opening a Tank Hatch, https://ehsdailyadvisor.blr.com/2016/02/what-you-can-do-to-avoid-vapor-hazards-when-opening-a-tank-hatch/

Sam Carbis Solution Group, LLC, Top 3 Tank Truck Safety Hazards: Do You Know The Risks? https://www.carbissolutions.com/blog/top-3-tank-truck-safety-hazards-do-you-know-the-risks/

Truckers Report Forum, New To Tankers – Safe Access to the Top? https://www.thetruckersreport.com/truckingindustryforum/threads/new-to-tankers-safe-access-to-the-top.1253136/

Additional Resources

SafeRack, Tanker Truck Loading, Truck Safety Cage and Fall Protection, https://www.saferack.com/product/loading-safety-cages/truck-safety-cages/

Rigid Lifelines is a division of SPANCO, Inc. Fall Protection for Tanker Trucks, Loading Docks, Flatbed Trucks, and Railcars, https://www.rigidlifelines.com/blog/fall-protection-for-tanker-trucks-loading-docks-flatbed-trucks-and-railcars/

FallProof, Transportation and Trucking Fall Protection: Implementing It Effectively, https://www.fallproof.com/blog/transportation-and-trucking-fall-protection-implementing-it-effectively/

Additional Resources: Other Hydrogen Sulfide Incidents Outside of the United States

11 dead after toxic gas leak in India, Chemistry World, https://www.chemistryworld.com/news/11-dead-after-toxic-gas-leak-in-india/4017404.article

Journal of Occupational and Environmental Hygiene: "Confined Space Hazards: Plain Seawater, an Insidious Source of Hydrogen Sulfide" (June 2023). Fatal Hydrogen Sulfide Incident in a Shipboard Confined Space, https://synergist.aiha.org/202309-hydrogen-sulfide-confined-space

Acute Communicable Disease Control (ACDC), 2005 Special Reports, DEATHS AND ILLNESS FROM HYDROGEN SULFIDE AMONG SHIP WORKERS, LOS ANGELES, CALIFORNIA, http://publichealth.lacounty.gov/acd/reports/spclrpts/spcrpt05/DeathsHydrogenSulfide05.pdf

Incident 10: Paper Mill Workers Exposed to a High Concentration of Hydrogen Sulfide That Was Venting from a Storage Tank That Contained Process Water (2 Fatalities, 14 Injuries, Including 1 Worker Placed into an Induced Coma Due to the Seriousness of His Injuries)

Norske Skog Paper Mill
Ettamogah (Near Albury), NSW
Incident date: May 24, 2018

A very serious venting of hydrogen sulfide gas occurred at the Norske Skog Paper Mill on May 24, 2018, overcoming the mill workers, killing two and injuring many others. Two paper mill workers were killed following exposure to hydrogen sulfide, and 14 were hospitalized. One maintenance contractor suffered hydrogen sulfide poisoning and hypoxic respiratory failure and was placed into an induced coma due to the seriousness of his injuries. He spent two weeks in hospital. The mill was shut down after the incident.

Hydrogen sulfide is commonly found in petroleum refineries, sewer treatment plants, and biological waste sites. It is also commonly found in paper mills due to the biological degradation processes in the paper manufacturing processes. Albury Mayor Kevin Mack told the local media, "The mill had a very good safety record." "The pulp itself emits the toxic gas we're talking about," Mr. Mack said.

In the pulp and paper industry, hydrogen sulfide may be generated during the heating processes, acid cleaning of processing equipment, or mixing acids with process liquids. Acid mixing with sulfur-bearing liquids can generate significant amounts of gas, including hydrogen sulfide. Hydrogen sulfide is also produced when the wood chips are mixed with white liquor and cooked at high temperatures. The heat breaks the bond of the glue holding the wood chips together and creates a chemical reaction, releasing hydrogen sulfide and mercaptans, a type of sulfur. Hydrogen sulfide can also be generated in the effluent ponds when acid and sour water mix.

The two workers died when they inhaled a high concentration of hydrogen sulfide that was venting from a storage tank that contained process water. Reportedly, there were uncovered openings in the tank, which allowed the vapors to be vented into the atmosphere, where they overcame the workers. These findings were published in a safety alert issued by SafeWork NSW titled "Exposure of workers to hydrogen sulfide gas."

Norske Skog is part of an international network of paper companies and was prosecuted by SafeWork NSW in District Court. The company pleaded guilty to failing to comply with the WHS Act for exposing workers to a risk of serious injury or death. The company was ordered to pay US$1.01 million. They were also ordered to create a process safety video to help educate the industry on these hazards. The penalty had been reduced due to the company stepping up and pleading guilty. Judge David Russell was quoted in the local media as saying, "Norske Skog's level of culpability was in the high range of seriousness and that the injuries and emotional harm as a consequence of the incident were substantial." He added, "In summarizing these Victim Impact Statements, I cannot possibly convey the searing pain and loss experienced by each family member, which was expressed in court as the statements were read out."

According to the Bureau of Labor Statistics, in the United States alone, during the seven years between 2011 and 2017, 46 workers died due to exposure to hydrogen sulfide. This makes H_2S one of the leading causes of workplace gas inhalation deaths in the country. Most of these occurred in the oil and gas industry. However, others occurred across various industries and work sites, including paper mills, sewer treatment plants, and waste disposal sites. Hydrogen sulfide is deadly in concentrations around 800 ppm and above, and the victim does not get a second chance.

Incident 10: Key Lessons Learned

The key lesson learned in this incident is that hydrogen sulfide is a serious hazard in petroleum refining and petrochemical plants but is equally hazardous in paper mills, sewer treatment plants, waste disposal sites, etc. This incident report does not clearly state a couple of details, which are discussed below.

In the paper mills, the H_2S is generated during the "cooking" phase and in some bleaching operations, especially if the "black water" encounters acid. H_2S in paper mills is just as deadly as in petroleum refining operations. Most US paper mills are equipped with fixed H_2S detectors and personal monitors worn by the employees and contractors working at the facility. It is unclear from this incident report or media reports that this facility provided this capability. If it existed, the devices certainly should have provided advanced warning of the presence of hydrogen sulfide, giving the employees the opportunity to escape. From my perspective, these devices certainly should have been provided. The personal monitoring devices available today include four-gas monitors with audible, visual, and vibration capabilities and are available for various gases.

I know that when these devices are periodically tested and properly worn, they save lives in refining and petrochemicals. I am equally convinced that personal gas detectors and fixed detectors can save lives in paper mills and other facilities that process hydrogen sulfide.

I have a second question concerning this incident. We know that there were 2 fatalities and 14 injuries in this event, including one injured person who spent a couple of weeks in the hospital. For many other incidents involving H_2S, when one person is exposed and collapses, a second person often rushes in to provide aid or to rescue the first victim. Many times, this results not only in one victim but also the attempted rescuer often also becomes a victim and dies from exposure to high-concentration hydrogen sulfide.

I can't help but wonder if that is what happened in this case. We may never know, but the key lesson learned is that if this happens at our site, we must fall back on our training and resist the human urge to render assistance. We should never rush in unless we are a trained rescuer and have immediate access to emergency response equipment, such as an SCBA and other personnel rescue equipment. Otherwise, one victim can easily turn into two, and sometimes even more. We must resist the temptation, immediately call for the trained rescue team, and then stand by and direct the rescuers to the site.

Additional Resources

Print21, Uncovered openings' at fatal gas leak site, https://www.print21.com.au/news/uncovered-openings-at-fatal-gas-leak-site

Trintias Group, $1M fine for Victorian paper mill over the death of two workers, https://trinitasgroup.com.au/2020/10/08/1m-fine-for-victorian-paper-mill-over-the-death-of-two-workers/

Australian Broadcasting Company News, Albury paper mill gas leak kills two Norske Skog workers, third in critical condition, https://www.abc.net.au/news/2018-05-25/albury-paper-mill-toxic-gas-leak-kills-workers/9798744

ISS Source, Aussie Paper Mill Gas Leak Kills 2, https://www.isssource.com/aussie-paper-mill-gas-leak-kills-2/

General Carbon Corporation, Pulp & Paper Mills, https://generalcarbon.com/industries-served/pulp-paper-mills/

Emerson, Hydrogen Sulfide in Paper Mills: Find it Before it Finds You, https://www.emersonautomationexperts.com/2020/control-safety-systems/hydrogen-sulfide-in-paper-mills-find-it-before-it-finds-you/

Industrial Scientific, What You Need to Know About Gas Hazards in Pulp and Paper Mills, What You Need to Know About Gas Hazards in Pulp and Paper Mills

Incident 11: Two Workers Were Killed While Inspecting Cargo Tanks on a Marine Vessel Due to Hydrogen Sulfide in an Innocuous Cargo

Danish chemical/product tanker NORD MAGIC
While at anchor off Chittagong, Bangladesh
Marine accident report on occupational accident
June 17, 2022 (2 Fatalities)

The following incident is based on articles from *The Journal of Occupational and Environmental Hygiene*, the National Library of Medicine, and the National Center for Biotechnology Information. Please see references and citations in the Additional Resources Section of this chapter. A detailed investigation report completed by the Danish Maritime Accident Investigation Board (DMAIB) is also available via a link in the Additional Resources section of this chapter.

This incident is exactly what is described in detail in Chapter 2, which describes how sulfate-reducing bacteria (SRBs), especially when there is little oxygen available and the conditions are anaerobic, actively consume sulfate, generating sulfide or the base chemical for hydrogen sulfide. The sulfide then reacts with available hydrogen from the water, resulting in the generation of hydrogen sulfide. The investigation of this incident clearly shows that this biological process can generate significant amounts of hydrogen sulfide and result in the death of two workers.

In 2022, two technicians died while conducting an inspection inside an empty cargo tank on a Danish tanker. The deaths were attributed to hydrogen sulfide poisoning due to the accumulation of H_2S inside the cargo tank while the tanker was sitting idle. The tanker's most recent cargo had been vegetable cooking oil, an innocuous cargo. This incident was described as enigmatic in one of the articles, which means that it was difficult to interpret or understand. I'm sure this was certainly the case; two deaths while inside a cargo tank that had previously contained vegetable cooking oil was a mystery.

What was found during the investigation was that the cargo tank had been washed with 10 cubic meters (about 2,600 US gallons) of seawater, and the seawater had lied dormant inside the cargo tank for 19 days before the workers entered the cargo tank, a confined space. No one considered the wash water to be a hazard to the workers. Also, gas tests that were made in the cargo space AFTER the deaths were measured at 100 ppm of hydrogen sulfide, which is IDLH for H_2S but well below that required to result in the worker's deaths.

As it turned out, when laboratory analysis was completed on the seawater, it was confirmed that the seawater contained enough sulfate to create an immediately fatal concentration of hydrogen sulfide in the 4,500 cubic meter tank (about 159,000 cubic feet). In fact, the calculation indicated that if the sulfate was mixed well, and all the sulfates had been converted to hydrogen sulfide, that the expected concentration of H_2S would be about 1,600 ppm, well above the amount that could take the lives of the workers. Also, a laboratory analysis of the wash water indicated that the hydrogen sulfide concentration may have been as much as 2,000 ppm. The 100 ppm, as measured immediately after the accident, was the maximum reading of the instrument used and did not indicate the actual H_2S concentration in the cargo tank.

The reports in the two journals mentioned that asphyxiation incidents of this type are "stubborn" and that strict ventilation requirements and accurate vapor space gas testing offer simple and effective preventative measures. Indeed, in the case of this incident, the Danish Maritime Accident Investigation Board (DMAIB) clearly said that the workers entered the tank "even though the tank had not been ventilated, the atmosphere had not been tested, and no work permit had been issued." These cardinal rules must be followed for every critical task, including confined space entry. Gas testing before vessel entry should always be done and include testing for oxygen (minimum 19.5%), flammables (maximum of 10% LEL), hydrogen sulfide (maximum of 20 ppm; 20% for OHSA, 10% for NIOSH), and other toxins that may be present based on the process involved.

Additional Resources

The Journal of Occupational and Environmental Hygiene, Hedlund, F. H. (2023). Confined space hazards: Plain seawater, an insidious source of hydrogen sulfide. *J Occup Environ Hyg*, *20*(8), 322–328. https://doi.org/10.1080/15459624.2023.2211638, https://oeh.tandfonline.com/doi/full/10.1080/15459624.2023.2211638

The Danish Maritime Accident Investigation Board (DMAIB), Marine accident report on occupational accident 17 JUNE 2022, https://dmaib.com/Media/638088635560435483/NORD_MAGIC_Occupational_accident_on_17_June_2022.pdf

Incident 12: One Person Was Killed by Hydrogen Sulfide During a Rescue Attempt While Performing Maintenance on Wastewater Pump in an Underground Confined Space

The Great Western Sugar Company
Fort Morgan, Colorado
May 29, 2024 (One person was killed during a rescue attempt, and 2 firefighters and 1 worker were treated and released at the hospital)

The next incident involving hydrogen sulfide involves what I believe to represent around 30–40% of the hydrogen sulfide deaths worldwide. I have looked at a lot of these tragic events and have concluded at least this number of deaths due to hydrogen sulfide exposure involves someone, more than likely a coworker, who has witnessed a teammate being overcome by H_2S. In these events, the worker is more likely than not overcome by the human urge to attempt a rescue when witnessing a coworker collapse. You have already seen one such incident that happened at a waste management site near Superior, Wisconsin, where four workers were killed, including two brothers, while working on a wastewater pump in an excavation at an industrial landfill site, three of whom were attempting rescue.

Attempted rescue is not planned. The person attempting the rescue is generally not trained in the required techniques, and in most cases, they are not equipped with the highly specialized equipment to effect a successful rescue.

A direct report from the US Occupational Safety and Health Administration on this fatal incident at the Great Western Sugar Company:

"At 3:22 pm on May 29, 2024, Employee #1 attempted to rescue a fellow employee who had been overcome by hydrogen sulfide (H_2S) gas in an underground pump house. Prior to this incident, multiple coworkers had been working in the pump house. They were exposed to H_2S gas emitted from a pipe containing water with by-products from sugar beets, dirt, mud, and chemicals used in water treatment. The victim became overwhelmed by the H_2S gas, and Employee #1 was also overexposed while attempting the rescue. The employee lost consciousness and was later found unresponsive."

The following is assembled from the OSHA report and as reported by local media:

The Great Western Sugar Company was founded in 1900 and processes sugar beets in Fort Morgan, Colorado, from the Fort Morgan area and Scottsbluff, Nebraska. Great Western was sold to a cooperative in the city of Fort Morgan, where beet processing is an important industry. The main building exterior is basically an original iconic structure, with several outbuildings and even underground structures containing processing equipment.

This incident occurred in one of these underground structures where maintenance workers performed routine maintenance on pumps that were part of the processing equipment. The employees performed routine maintenance at a utility building detached from the main plant in Fort Morgan when they were "unexpectedly exposed" to "naturally occurring" hydrogen sulfide fumes. The pumps were approximately 10 feet below ground in what would be classified as a confined space. One of the reports indicated that the pump house was 20–30 feet underground; this has not been confirmed.

The pumps were part of the wastewater treatment plant, which was only 2½ years old. As they opened the seal, sludge began to leak out of the pumps, releasing hydrogen sulfide into the confined space. The workers were reportedly wearing protective equipment, but as the local County Sheriff reported, one of the workers became combative and dislodged his breathing apparatus.

When this incident occurred, the first responders were called due to one person being unconscious. While enroute to the incident location, the first responders learned a second person was also unconscious. Several responders from the local community arrived just after 3:30 p.m., and when they first arrived, they were advised that this was a possible ammonia leak. However, the rescue teams quickly determined through air monitoring that the release was hydrogen sulfide. After appraising the scene, they called for additional assistance from the Brush Volunteer Fire Department for additional manpower to aid the confined space rescue and potential decontamination. Firefighters entered the confined space wearing hazmat PPE, including SCBA; however, they were also exposed while attempting to rescue one of the workers.

The rescue operation lasted about 70 minutes, according to reports. Three employees and two firefighters were quickly transported to Saint Elizabeth Hospital in Fort Morgan, where they were treated and released. Unfortunately, the 33-year-old employee who attempted the rescue was pronounced deceased, and another employee remained hospitalized the following morning.

Incident 12: Key Lessons Learned

Hydrogen sulfide is a naturally occurring by-product of the production's wastewater and escaped when the workers opened the seal on one of the wastewater treatment pumps, which were in a confined space underground.

Hydrogen sulfide is a flammable, colorless gas that smells like rotten eggs at low concentrations, but at higher concentrations the H_2S deadens the sense of smell, and the victim may believe it has dissipated. Hydrogen sulfide occurs naturally in fossil fuels and from the biological breakdown of organic matter. According to the US Environmental Protection Agency and the US OSHA, a few breaths of the toxin can be deadly.

In this incident, the workers entered a confined space. While they were wearing respiratory equipment, it is not clear from the reports that all other requirements of the OSHA Confined Space Entry Standard (OSHA 29CFR 1910.146) were being followed. These include the following (note: this is a summary; please refer to the standard for additional details):

- **Workplace Assessment and Evaluation:**
  All US employers must conduct a workplace assessment to identify any spaces that meet the definition of permit-required confined spaces. This includes spaces that may contain hazardous atmospheres, potential engulfment hazards, or internal configurations that could become a trap for workers.
- **Documented Confined Space Program:**
  All US employers must have a comprehensive written program that includes the procedure for hazard identification and control. This includes training for personnel who may be authorized to enter a confined space or serve as attendants and supervisors.
- **Confined Space Entry Permit System:**
  All employers must have a written confined space entry permit detailing the specific space, potential hazards, entry procedures, and personnel involved in the confined space entry process. These procedures must be used for every permit-required confined space entry.
- **Monitoring the Work Atmosphere:**
  The workspace must be tested for oxygen first, followed by testing for flammable gas and toxic substances. In almost all cases, the toxic gas test will test for hydrogen sulfide due to its possibility of being naturally occurring, followed by other toxic substances based on the product(s) that have been processed in the equipment.

  Continuous monitoring may be required if warranted by the material involved or other work conditions. If continuous monitoring is not done, the gas testing should be completed within two hours of work commencing.
- **Standby Attendant:**
  The confined space entry procedure should require a standby attendant to be stationed just outside the confined space and in communication with all entrants. If needed, the attendant should be prepared to activate the rescue plan or medical response.
- **Rescue Plan/Medical Response:**
  The confined space entry procedure must also include a means of immediately rescuing entrants in an emergency by a trained and capable team with ready access to rescue equipment and appropriate PPE.

Trained medical personnel should also be available on call in the event a need arises for the entrants or other personnel involved in confined space entry.

- **Training:**
  All personnel involved in confined space entry must be adequately trained in specific hazards, procedures, and emergency response protocols.
- **Personal Protective Equipment:**
  All personnel involved in confined space entry in the roles of entrant, standby attendant, or rescue shall have access to appropriate PPE. This typically includes an SCBA and fall protection, as required by the hazards involved. The rescue team members may require specialized rescue equipment, including extraction equipment.
- **Confined Space Purging/Ventilation:**
  As determined by the gas test results, if atmospheric purging or continuous ventilation is necessary, it must be used to eliminate or abate the possibility of a hazardous atmosphere. This must be completed before entrants enter the confined space and continued at all times while personnel are there.

  Typically, ventilation is accomplished using a mechanical fan or blower with air ducts to ensure the forced air is properly routed to the entrants. When used in areas where flammable vapors may be present, the motor driver must be approved for use in electrically classified areas (Class 1, Division 1). This means that the electrical equipment is designed to contain an internal explosion and prevent electrical energy from causing the ignition of surrounding vapors. The equipment and wiring are also sealed to prevent outside vapors from entering the equipment.

Death During Attempted Rescue:

The bottom line in this incident, which led to the fatality, was the worker witnessed his coworker collapse due to encountering hydrogen sulfide and immediately attempted to rescue him. Unfortunately, it was reported that his SCBA was dislodged during the rescue attempt, and he was also overcome and killed by the toxic nature of the deadly gas.

Science Direct and Forensic Science International reported how six workers died in an unusual case of hydrogen sulfide poisoning. This was reported in an article dated March 2016 in an article titled "Domino effect: An unusual case of six fatal hydrogen sulfide poisonings in quick succession." All six workers were overcome by hydrogen sulfide while unblocking a wastewater cistern in Italy. The first worker cleared the blockage and then died within minutes due to H_2S exposure. Afterward, the other five workers died when they entered the excavation one by one to provide aid to their coworkers. The hydrogen sulfide killed all six workers. Again, this clearly illustrates how deadly this toxic gas is and how quickly it can lead to death.

The Bottom Line:

This comes down to training. If anyone reading these case studies should witness a coworker collapse, please respond and attempt a rescue, but only if you have been extensively trained in rescuing others who are in trouble and have all the appropriate rescue equipment. This is especially true for workers in oil and gas operations or any sewer, wastewater, manure storage, or treatment operations where hydrogen sulfide is possible.

Otherwise, most of us should immediately leave the area and call for the trained professionals, then standby to direct them to the victim as soon as they arrive. As I said at the beginning of this case study, this happens all too often, and unfortunately, the attempted rescuer(s) also become a statistic.

Believe me, this is hard guidance to give you. I understand that none of us want to see our coworker or our buddy die. But it is truly a sad day when someone is killed attempting to rescue their coworker.

Additional Resources

US Occupational Health and Safety Administration (OSHA), Investigation Summary: Investigation Nr: 167012.015, Employee dies when exposed to H_2S gas while attempting rescue, https://www.osha.gov/ords/imis/establishment.inspection_detail?id=1751705.015

CBS News Colorado, 1 dead, 4 injured after possible chemical leak at sugar plant in northeastern Colorado, https://www.cbsnews.com/colorado/news/4-hospitalized-possible-chemical-leak-sugar-plant-northeastern-colorado/

Domino effect: An unusual case of six fatal hydrogen sulfide poisonings in quick succession, Science Direct and Forensic Science International, https://www.sciencedirect.com/science/article/abs/pii/S0379073816000402

California Acute Communicable Disease Control; 2005 Special Reports, DEATHS AND ILLNESS FROM HYDROGEN SULFIDE AMONG SHIP WORKERS, LOS ANGELES, CALIFORNIA, http://publichealth.lacounty.gov/acd/reports/spclrpts/spcrpt05/DeathsHydrogenSulfide05.pdf

Chapter 7: End-of-Chapter Quiz (Note Some Answers Are Found in Other Chapters of This Book)

1. Please name at least four process safety systems and facilities that have been significantly improved over the past twenty-plus years and should have reduced the number of hydrogen sulfide fatalities over this same period.

2. Please provide one example that indicates that lives continue to be lost due to loss of containment of the highly toxic hydrogen sulfide gas into the workplace.

3. About how high can the hydrogen sulfide concentration be in a typical petroleum refinery hydrogen sulfide treating system?

4. How does 80–90% hydrogen sulfide convert to parts per million?

5. What is the potential outcome if two incompatible chemicals are inadvertently mixed, for example, unloading the wrong chemical into a chemical storage tank?

6. What atmospheric gas testing should be completed before entry into any process vessel, storage tank, tank truck or tank car, or any excavation 4 feet or more in depth?

7. What is the definition for an OSHA permit-required confined space? Provide definition as per OSHA 29CFR 1910.146:

8. List six examples of enclosures that meet the OSHA definition of a permit-required confined space.

9. Approximately 15–20% of all fatalities involving hydrogen sulfide result in more than one fatality; occasionally, it involves as many as three or four fatalities. Why is this the case?

10 When a truck driver is overcome by H_2S while sampling or gauging a tank truck, they are frequently killed by falling from the truck. Is the cause of an incident like this the failure to recognize the H_2S hazard, or is it failure to wear fall protection?

11. When performing maintenance on a process system with a high concentration of H_2S, what additional personal protective equipment should be worn by all craftsmen?

12. What is one of the most important personal safety protective devices that workers should wear anytime they are working around or near equipment containing high concentrations of hydrogen sulfide?

13. When the H_2S detector alarm sounds, how much time do we have before we should leave the area?

14. When unloading a truck of chemicals, the following should be in place to prevent from unloading chemicals into the wrong tank.
    Please mark all that apply:
    a. The facility unloading lines should be correctly labeled, color-coded, and identified for the specific chemical to be unloaded.
    b. Always verify the shipping documents that are provided by the truck driver to ensure the correct chemical is being delivered.
    c. The truck driver should know what chemical is being delivered.
    d. The unloading connections should be designed only for the chemical to be unloaded and incompatible with all other chemicals.
    e. The truck driver and the facility representative, typically the operator, should both be present during the unloading.
    f. The company should have a documented procedure for unloading the chemical and this procedure should be followed.
    g. All of the above.

15. When gas testing a vessel for entry, why is it important to test for oxygen before testing for flammables or hydrogen sulfide?

16. When considering an excavation for a confined space, what depth triggers all the permit-required confined space entry requirements?

17. There are many requirements detailed by OSHA 29CFR 1910.146, the Confined Space Entry Standard. What are the three of the most important elements of the OSHA Confined Space Entry Standard?

18. Please select the most likely places where serious levels of hydrogen sulfide may be expected and where precautions, including the use of self-contained breathing apparatus (SCBA), should be required.
    Please mark all that apply:
    a. A grain silo with rotting grain in the bottom with sludge and water.
    b. A manure pit where manure from a cattle farm has been dumped.
    c. An old and abandoned oil and gas well with leaky pipes near the surface.

d. A tank truck hauling water bottoms for disposal from crude oil storage tanks.
e. A settling pond at a municipal sewerage treatment plant.
f. A volcanic geyser in a national park.
g. An open marshy field with decaying grass and other vegetation.
h. A sewer manhole where underground sewerage transfer pumps are located.
i. An empty railcar that recently carried a load of crude oil.
j. A pit containing a pump that removes water draining from an adjacent landfill.
k. None of the above.
l. All of the above.

19. Your task is to add polymer to an asphalt tank through a gauging hatch on the top of the storage tank. What is the safest way to approach the gauging hatch?
a. Approach the gauging hatch from the downwind direction.
b. Approach the gauging hatch from the crosswind side of the hatch.
c. Don a self-contained breathing apparatus (SCBA) and approach the hatch from the crosswind side of the hatch.
d. Don a self-contained breathing apparatus (SCBA) and approach the hatch from the upwind side of the hatch.

20. You are working with a crew that is using a hydro jet system to unclog the Quick Trip sewer line. After the crew successfully dislodged the obstruction, you are requested to enter the manhole to visually inspect the condition of the line. The manhole is eight feet deep. How do you prepare to enter the manhole?
Please mark the most correct answer.
a. Gas test for oxygen and flammables, and if the tests are good, enter the manhole and perform the inspection.
b. Ensure a gas test for oxygen, flammables, and hydrogen sulfide. Prepare a confined space entry permit documenting the gas test results, including a rescue plan, and a standby attendant.
c. Prepare a confined space entry permit including a gas test for oxygen, flammables, and hydrogen sulfide. Prepare a rescue plan, including wearing a safety harness with a retrieval rope attached. Ensure a standby attendant who can notify rescuers.
d. Don a self-contained breathing apparatus (SCBA) and enter the manhole and perform the inspection.

8

The Hierarchy of Controls

Introduction

The National Safety Council introduced the Hierarchy of Controls in 1950 as a safety system to reduce workplace hazards and on-the-job injuries. This was quickly seen as an effective process and was incorporated into US Government regulations when the US Occupational Safety and Health Administration (OSHA) was formed in 1970.

The Hierarchy of Controls got a significant boost when professor and author Trevor Kletz, in 1977, spoke to the Society of Chemical Industry in England on "What You Don't Have, Can't Leak," which obviously focused on the issue of hazard elimination. Kletz suggested that, where it is feasible, risk should be managed by hazard elimination rather than providing safety systems and personal protective equipment.

This evolved into Inherently Safer Design, and Dr. Kletz became the industry's recognized grandfather of Inherently Safer Design. The Hierarchy of Controls and Inherently Safer Design are well-known throughout the Process Safety discipline and are taught at most engineering universities worldwide. The Hierarchy of Controls will be described in this chapter.

References for both the Hierarchy of Controls and Inherently Safer Design will be included in the Additional Resources section of this chapter.

The Hierarchy of Controls

US OSHA states the following about the Hierarchy of Controls: "The hierarchy of controls is a method of identifying and ranking safeguards to protect workers from hazards. They are arranged from the most to least effective and

Hydrogen Sulfide (H_2S) Hazards: Risk Recognition and Prevention, First Edition. M. Darryl Yoes.

include elimination, substitution, engineering controls, administrative controls and personal protective equipment." They go on to say, "Often, you'll need to combine control methods to best protect workers. For example, a local exhaust system (an engineering control) requires training, periodic inspections, and preventive maintenance (administrative controls). You will also need to consider feasibility."

OSHA presents the Hierarchy of Controls in the following order of importance:

- Hazard elimination (is always first)
- Substitution
- Engineering controls
- Administrative controls
- Personal protective equipment (PPE) (PPE is always the last line of defense)

NIOSH (The National Safety and Health Administration) emphasized the Hierarchy of Controls in the 1980s; however, it came into being in 2007 when NIOSH included it in the Prevention through Design program. This program included what became known as the Hierarchy of Controls, which is the five-step program we know today. The Hierarchy of Controls is illustrated in Figure 8.1.

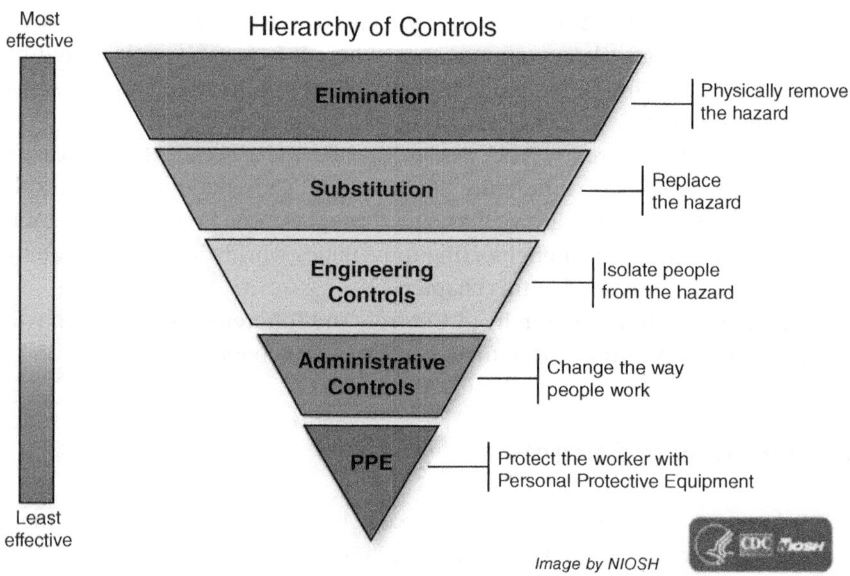

Figure 8.1 The Hierarchy of Controls. From About Hierarchy of Controls (NIOSH) (April 10, 2024). *Source:* The National Institute for Occupational Safety and Health (NIOSH)/ National Institute for Occupational Safety and Health (NIOSH)/Public Domain.

Please note that hazard elimination is by far the most effective, and that workers' protection by adding personal protective equipment (PPE) is the least effective. The concept is that if the hazard is not present, it won't harm the workers.

Let's break down the concept of the Hierarchy of Controls into the individual five key elements:

- Hazard Elimination:
  US OSHA defines hazard elimination as "removing the hazard completely." They provide the following examples: "Ending the use of a hazardous material. doing work at ground level rather than at heights or stopping the use of noisy processes."

  Real-life examples include converting a process that stores a toxic chemical in tanks or receiver vessels to a "use-on-demand" process where the excess inventory of a toxic chemical is eliminated. While this may also fit into the topic of Hazard Substitution, it reduces the amount of hazardous or toxic material subject to leakage or loss of containment.
- Substitution:
  US OSHA defines substitution as "replacing the hazard with something less dangerous or switching to a process that uses less force, speed, temperature, or electrical current."

  Real-life examples include converting the solvent used in a lube extraction process from the toxic chemical phenol to a less toxic chemical. This clearly is substituting a toxic chemical for a non-toxic one, thereby reducing the risk of personnel exposure or environmental incidents.
- Engineering Controls:
  Engineering controls can be used to reduce the opportunity for contact between the chemical or other hazards and the worker. This can be done by implementing protective barriers or guards or by improved ventilation to eliminate toxic vapors. NIOSH has developed a Controls Database that includes examples of engineering controls. Please refer to the Additional Resources section at the end of this chapter.
- Administrative Controls:
  With administrative controls, we literally change the way people perform work. Administrative controls may change or reduce the work shift to limit the exposure time, frequency, or intensity of the hazard.

  The administrative controls may provide additional job-related skills training or schedule job rotation, periodic rest breaks, limit access to hazardous areas, etc.
- Personal Protective Equipment (PPE):
  The use of PPE should be the last priority for protecting workers after Hazard Elimination, Substitution, Engineering Controls, and Administrative Controls.

However, it is extremely important that the workers are properly trained in the proper use of PPE, and that the proper PPE is worn correctly at any time the hazard is present.

Each employer should have a PPE Program that clearly spells out when PPE is required and what type of PPE should be used. Examples of PPE include protective gloves, safety glasses (generally with side shields), hearing protection, hard hats, and respirators, including self-contained breathing air respirators when needed.

The employer's PPE Program should include the following at a minimum:

- Workplace hazards assessment
- PPE selection and use
- Inspection and replacement of damaged or worn-out PPE
- Employee training

The two most difficult elements to implement and administer are administrative controls and PPE. These require a significant upfront and ongoing effort by the workers and managers to get the programs in place and to keep them going. However, they can be very helpful when the employers implement other control elements in the Hierarchy of Controls. Administrative controls and PPE are also typically applied to existing operations, especially when the hazards are not well identified, or other controls are still in the implementation phase.

Applying Hierarchy of Controls to Hydrogen Sulfide

Applying the first two elements of the Hierarchy of Controls (Elimination and Substitution) to hydrogen sulfide becomes difficult or almost impossible. Since H_2S is naturally occurring and an inherent hazard in petroleum processing, it can't be replaced or substituted, at least not with current technology. Having said this, some measures can be taken, such as selecting a different, less sour grade of crude oil, which significantly reduces the H_2S concentration in downstream units. Of course, this comes with an expensive cost component, making it far less attractive.

However, with the remaining three elements, we do have a significant degree of control and can make a significant impact in improving safety in operations, which should always be our objective.

Engineering Controls: We can implement engineering controls to help isolate workers from H_2S or minimize their exposure. For example, we can implement improved ventilation with exhaust fans to remove H_2S from the workspace or dilute its concentration. We can ensure that all vent systems are routed to the flare and not to the atmosphere and use improved treatment systems to reduce or eliminate H_2S from product streams. When collecting samples, sample points for sour products should be engineered to protect workers from H_2S vapor.

Administrative Controls: We can ensure administrative controls such as improved operations procedures with step-by-step work sequencing, adequate cautions and warnings, and hazard analysis exercises and training to reduce the likelihood of H_2S exposure. Ensure that job hazard analysis (JSA) is routinely conducted to identify job hazards and improvement actions for specific tasks. Ensure detailed and thorough work permits with required gas testing to control access to all areas where H_2S may be present.

Personal Protective Equipment (PPE): And, always as the last resort, ensure the correct PPE is readily available and that the workers are using it properly when working in H_2S areas. Ensure that requirements are in place for H_2S detectors to be worn anytime personnel enter the areas where H_2S may be present, that this requirement is enforced, and that they immediately leave the area when it alarms. Ensure that the self-contained breathing apparatus is worn when process equipment is opened and at all times when potential exposure to H_2S exists.

Additional Resources: The Hierarchy of Controls

US Occupational Safety and Health Administration (OSHA), Identifying Hazard Control Options: The Hierarchy of Controls, https://www.osha.gov/sites/default/files/Hierarchy_of_Controls_02.01.23_form_508_2.pdf

The National Institute for Occupational Safety and Health (NIOSH), The Hierarchy of Controls, https://www.cdc.gov/niosh/hierarchy-of-controls/about/index.html

Texas Department of Insurance, Reducing hazards with the Hierarchy of Controls, https://www.tdi.texas.gov/tips/safety/hierarchy-of-controls.html#:~:text=In%201950%2C%20the%20National%20Safety,OSHA)%20was%20formed%20in%201970

The National Institute for Occupational Safety and Health (NIOSH), Engineering Controls Reports, https://www.cdc.gov/niosh/engcontrols/ecd/default.html

The National Institute for Occupational Safety and Health (NIOSH), Directory of Engineering Controls, https://www.cdc.gov/niosh/engcontrols/default.html

Canadian Centre for Occupational Health and Safety, Hazard and Risk – Hierarchy of Controls, https://www.ccohs.ca/oshanswers/hsprograms/hazard/hierarchy_controls.html

Safety + Health Magazine, The Hierarchy of Controls, Strategy for safety singles out hazards before work starts, https://www.safetyandhealthmagazine.com/articles/16790-the-hierarchy-of-controls

US Occupational Safety and Health Administration (OSHA), A PRODUCT OF OSHA'S RECOMMENDED PRACTICES FOR SAFETY & HEALTH PROGRAMS, Identifying Hazard Control Options: The Hierarchy of Controls, https://acrobat.adobe.com/id/urn:aaid:sc:VA6C2:30f71883-7b7f-4728-a0b7-0eac380adf9f

Literature Cited

Kletz, T.A. (May 6, 1977). What you don't have, can't leak. *Chemistry and Industry*. pp 287–292.

Additional Resources: Inherently Safer Design

Dennis Hendershot; The Center for Chemical Process Safety, Inherently Safer Design, The Fundamentals, https://www.aiche.org/sites/default/files/cep/20120140-1.pdf

The Chemical Engineer, Trevor Kletz – A lifetime spent saving lives, https://www.thechemicalengineer.com/features/cewctw-trevor-kletz-a-lifetime-spent-saving-lives/

Dr. Trevor Kletz, https://www.aiche.org/bio/dr-trevor-kletz

Human Factors 101, Remembering Trevor Kletz, https://humanfactors101.com/2020/10/31/remembering-trevor-kletz/

Purdue University, Implementing Inherent Safety, https://engineering.purdue.edu/P2SAC/presentations/documents/David%20Moore-AcuTech-Implementing-Inherent-Safety-P2SAC2019-4-Dec-19.pdf

Chapter 8: End-of-Chapter Quiz

1. The Hierarchy of Controls was introduced by the National Safety Council in __________ as a safety system to reduce workplace __________ and reduce on-the-job injuries.

2. US OSHA states the following about the Hierarchy of Controls, "The hierarchy of controls is a method of __________ and __________ safeguards to protect workers from hazards."

3. How is the Hierarchy of Controls arranged?
   Please choose the most correct response.
   a. They are not arranged in any particular way.
   b. They are arranged in order from the least effective to the most effective.
   c. They are arranged in order from the most effective to the least effective.
   d. It depends entirely on the hazard being evaluated as to how they are arranged.

4. Please name the five controls in their appropriate order.

5. Which of the Hierarchy of Controls is the most effective?

6. Which of the Hierarchy of Controls is the least effective?

7. What are the factors that make PPE the least effective control?

8. Likewise, what makes Elimination the most effective of all the controls?

9. Which two of the controls are very difficult to implement in an existing process?

10. Which three of the controls can we easily make safety changes or improvements to address identified safety deficiencies or hazards?

9

Controlling the Hazards of Hydrogen Sulfide

Background

As presented in this book, hydrogen sulfide (H_2S) is as deadly as it has ever been, and lives are still being lost due to exposure to high concentrations of this deadly gas. Several incidents involving H_2S have been covered in the case studies presented in this book; however, this is only a very small percentage of the actual number of incidents happening in the United States and worldwide. I see this as a significant opportunity to save lives and prevent tragedy affecting families for a lifetime.

All too often, when one person succumbs to this deadly gas, someone else, such as a workmate, jumps in to render aid or to rescue the first person who collapses and is also killed by H_2S. In the case studies, I provided one example where the first person succumbed to H_2S. Then, in rapid succession, three more were killed as they entered the excavation one by one to help with the rescue, resulting in four deaths, including two brothers. In another case study, a father was overcome and died, and his son was killed while attempting rescue. Unfortunately, these stories have been repeated for decades, and unfortunately, they are still happening today. The point is that no matter if it is one fatality or several fatalities, there are too many, and we must be aware of the hazards of hydrogen sulfide, learn how to work with and around it safely, and ensure that our procedures and practices are fully aligned to prevent unexpected releases and protect personnel.

I recognize that this is easy to say but much harder to do. I have pointed out in this book that hydrogen sulfide is widespread and occurs naturally and in industrial production and manufacturing processes. It occurs naturally by decomposing human and animal waste and by biological degradation processes such as decaying vegetation. In production processes, sulfur compounds, including

Hydrogen Sulfide (H_2S) Hazards: Risk Recognition and Prevention, First Edition. M. Darryl Yoes.

hydrogen sulfide, occur naturally in crude oil and natural gas and are converted into super-concentrated hydrogen sulfide during the processes at petroleum refineries and some petrochemical plants. Hydrogen sulfide is also present in other industrial processes such as paper mills, municipal sewer treating, textile manufacturing, food processing, asphalt paving, mining, and tanning operations, to name a few (this is not intended to be a complete list).

Training

One place to start when we attempt to get this under control and stop, or at least limit, the number and consequence of H_2S exposures is with training and education. Hopefully, this book is a start. Personnel in the above industries must be trained in the hazards of H_2S, including where it should be expected at their facilities, how to detect the presence of H_2S, and how to protect themselves and others from the toxic nature of H_2S.

This means that each person working at the facility who may one day be at risk of H_2S exposure must participate in H_2S hazard awareness training, including what it is, the hazards associated with H_2S, and how they can protect themselves from the toxic nature of H_2S. Following the initial H_2S awareness training, a refresher course should be considered mandatory on a frequency not less than annually.

This may not be easy to do, but it must be done. This means everybody, including both company personnel and contractor workers, who may be in a position where they may be exposed, must receive both initial H_2S training and an annual refresher. This certainly means all production personnel, typically operations. Additionally, this includes truck drivers, laborers, maintenance workers, craftsmen, and others who may be exposed to the hazards of hydrogen sulfide.

This should be straightforward for most industries as they have an internal organizational structure to aid communication. Most industries also provide safety training and track the training to ensure it is done. However, it is harder for farms or dairy operations where the administration of these rules or policies is more difficult to communicate, and the training is not as well organized. I suggest these be done through farmer and dairy cooperative associations such as the Dairy Farmers of America (DFA), the American Farm Bureau, the National Cattleman's Beef Association, and other similar farm and dairy associations. This is important because H_2S is present in many of the day-to-day activities of the farmers and dairymen, and they are equally affected by its toxicity.

As I was writing this chapter, I ran across two very good articles written specifically about the toxic and flammable nature of the gas that can evolve from manure

Photo by Penn State Extension and John Tyson

Figure 9.1 A cattle manure transfer operation illustrating the hazards of managing the potential of H_2S.

and the hazards of working with the storage and handling of manure. Both articles were written by Penn State Extension and are posted on their website.

The first article, "Manure Gas Safety," written by John Tyson, discusses the four major gases that can evolve from manure. The four gases mentioned in the article are ammonia (NH_3), carbon dioxide (CO_2), methane (CH_4), and hydrogen sulfide (H_2S). He reported that although each gas has its own characteristics, hydrogen sulfide is the one that is involved most often when there is a fatality. There is a link to this article in the Additional Resources section, and I encourage each reader of this book to take the time to review this article. Figure 9.1 is an image of a manure transfer and illustrates the potential H_2S hazards that must be managed on a large livestock ranch or farm.

I'll talk about the second article when we discuss H_2S detection in the next paragraph of this chapter.

H_2S Detection Devices (Detectors and Monitors)

The second Penn State Extension article was written by Eileen E. Fabian (Wheeler), PhD, Michael Pate, PhD, and Dan Hofstetter, all of Pennsylvania State University, and is titled "Use Personal Gas Monitors to Avoid Exposure to Toxic Hydrogen Sulfide." This is also an extremely important article with a strong focus on the personal H_2S detector and the installation and use of fixed H_2S monitors.

These personal H_2S detection or monitoring devices go by both names. Some are referred to as detectors; some are referred to as monitors. They mean the same thing and have the same function. I have used the terms interchangeably in this book.

As I'm sure you know by now, after getting this far into this book, I am a strong believer in the value of H_2S detectors. I believe and support requiring everyone working in proximity to sources of H_2S to wear a personal H_2S detector. This article goes into detail on when they should be used and how to ensure that they are fully functional before they go to the field.

Similar gas monitors are also readily available for other gases; however, the most common monitor is for hydrogen sulfide (H_2S). The most common four-gas personal gas monitor provides continuous testing for oxygen (O_2), carbon monoxide (CO), hydrogen sulfide (H_2S), and combustibles (LEL). However, they can be designed for other toxic gases as well. These monitors typically have audibled, visual, and vibration alert warnings. Suppliers can meet your needs, whatever they may be, for the gas that represents a potential hazard at your facility. Figure 9.2 is an image of a personal hydrogen sulfide detector.

Whether you are working on a gas and oil production site in the Texas Permian Basin or in Alberta, Canada, on a large oil refinery unit in Louisiana or Oklahoma, a cattle ranch or dairy, or at a municipal sewer waste treatment plant, an H_2S detector can save your life. You will find a link to the Penn State

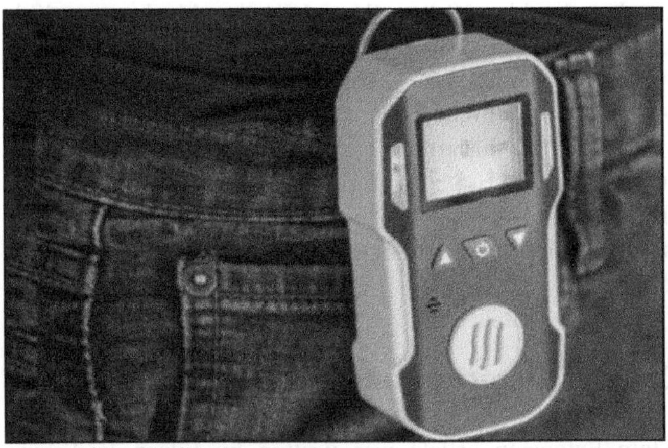

Courtesy of Forensics Detectors

Figure 9.2 A photograph of a personal H_2S monitor. Other types of toxic gas or oxygen monitors are available from Forensics Detectors and other safety suppliers. For example, Forensics Detectors provides detectors covering a wide range of toxic chemicals and oxygen deficiency, as well as a four-gas monitor (https://www.forensicsdetectors.com/blogs/articles/hydrogen-sulfide-monitor-manure).

article "Use Personal Gas Monitors to Avoid Exposure to Toxic Hydrogen Sulfide" in the Additional Resources section. I encourage you to download it and go through it in detail.

Fixed Hydrogen Sulfide Detectors

Personal hydrogen sulfide detectors are invaluable for operators and others who work in environments where H_2S may be present in high concentrations. However, fixed monitors can also provide early warnings of the presence of these dangerous and deadly gases. Fixed monitors should be installed on process units or work areas where people may be present during an unexpected release of hydrogen sulfide. These detectors provide continuous coverage 24 hours a day and use sensors sensitive to the gas. They are designed to detect low levels of hydrogen sulfide before it becomes toxic to personnel. Fixed H_2S detectors typically provide both visual and audible warnings. The visual is typically a flashing-colored light, and the audible may be a horn or a siren. Other types of warnings can also be designed into the system, such as sending alarms via text message or other notification systems.

These detectors are typically installed on process units in oil and gas refineries, near municipal wastewater treatment systems, sewer systems, and other industrial settings where hydrogen sulfide may be present. Some cattle and ranch operations have also installed fixed H_2S monitors near waste handling and disposal equipment.

The National Institute for Occupational Safety and Health (NIOSH) has established a recommended exposure limit (REL) of 10 parts per million (ppm). The Occupational Safety and Health Administration (OSHA) has established a permissible exposure limit (PEL) for hydrogen sulfide of 20 ppm over an 8-hour workday for general industry and 10 ppm for shipyards and construction. Most companies calibrate the fixed H_2S monitors to alarm before the hydrogen sulfide concentration exceeds 10 ppm REL. Depending on their internal philosophy or specific requirements specific to the company or industry, some companies may calibrate the alarms at even lower levels to provide workers and others with an early warning. As a reminder, it is important always to wear appropriate safety PPE anytime you are working in an H_2S area since even low levels of exposure can result in eye irritation, headache, and nausea. Also, exposure to low concentrations over extended periods can harm the olfactory nerves and affect the sense of smell.

The installation of the fixed detectors is only half the battle. Once the fixed detectors are installed, they must be maintained continuously. This means periodic testing, typically monthly, by verifying their operation with a calibration

Courtesy of US Chemical Safety and Hazard Investigation Board (CSB)/ Public Domain

Figure 9.3 Photograph of a hydrogen sulfide detector located at the Aghorn Waterflood Station near the pump house and control room building.

gas to ensure they are working correctly. You will note that at the Aghorn Water Flood Station incident, where two young people lost their lives, there was no indication that the fixed detection system was functioning correctly at the time of this incident. Unfortunately, some of the detectors were set to a testing mode, which prevented them from working correctly. These detectors are potentially life-saving devices and must receive proper attention, periodic testing, and maintenance. Also, the Aghorn management could not produce testing or maintenance records for the H_2S alarm system. Figure 9.3 is an image of a hydrogen sulfide detector.

Additional H_2S Hazards Associated with Ranch and Cattle Operations

Manure pits and storage areas such as tanks are periodically agitated to help aerate the mass and to help with the natural biological degradation process. Agitation also helps prevent the settling of solids, which can interfere with pumping the manure and reduce manure storage capacity. Agitation also helps reduce nutrient variability and stratification in the manure. However, manure agitation can also result in the release of toxic or flammable gases from the manure. These include hydrogen sulfide, ammonia, methane, and carbon dioxide. Hydrogen sulfide is the most dangerous gas because of its characteristics and hazards. Please see

Figure 9.4 for an H_2S warning sign in an area near manure agitation.

As previously discussed, H_2S can quickly deaden the sense of smell, making the workers believe that it has dissipated and is no longer a threat. At high concentrations, typically above about 1,000 ppm, death can occur in seconds. Hydrogen sulfide and methane are extremely flammable and explosive. Workers involved with manure agitation or who may be working nearby should always wear an H_2S detector and immediately leave the area if the alarm is activated.

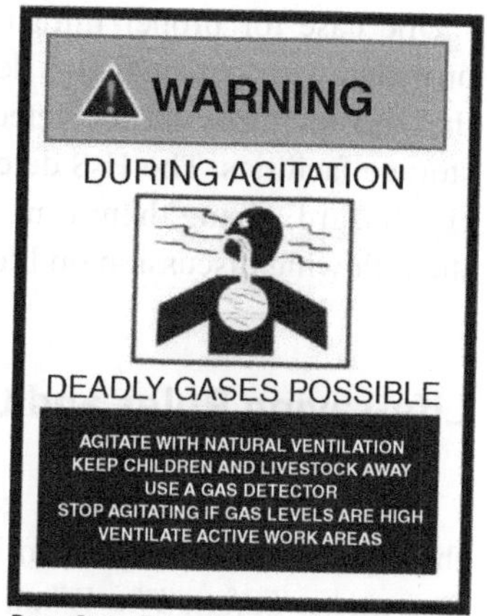

Penn State Extension and USDA NRCS (Pennsylvania)/Public domain

Figure 9.4 Image of a hydrogen sulfide warning sign located adjacent to an area where manure agitation may be in progress.

Risk Assessment and Personal Gas Detectors

A Simple Definition and Discussion of Risk Assessment

The definition of a risk assessment is the Probability of a defined scenario occurring multiplied by the Consequence if that defined scenario does occur. When we evaluate a potential scenario for risk, we evaluate the Potential by, for example, determining the number of times the scenario has occurred. The idea is that the more often the scenario occurs, the more likely there can be an incident; therefore, the Probability is higher.

Likewise, we also evaluate the Consequence, or the result, if the scenario does occur. For example, if the consequence of the scenario is a minor injury or the loss of containment of a nontoxic or nonflammable, the Consequence would be low. However, if the Consequence results in a significant loss of containment of a toxic chemical or a fatality, this would be evaluated as a high-consequence event.

In the evaluation, if either the Probability or the Consequence is high, additional mitigation would be required to reduce one or both to a lower level of risk.

Where risk assessment is applied to exposure to hydrogen sulfide, a defined scenario would easily be the loss of life of an operator working near equipment in high H_2S service without proper personal protective equipment (PPE).

For this scenario, unfortunately, we already have cases where this has happened. Operators have been killed by loss of containment of H_2S, and they occurred at a time when the operator was not wearing the proper PPE (detectors, SCBA, rescue equipment, etc.).

One case for proper mitigation is to ensure that operators (and others) are provided with personal H_2S detectors. A system should be established to ensure that the operators use the detectors anytime they are near high H_2S equipment or storage facilities. The H_2S detector would immediately warn the operator of the H_2S hazard, giving them time to evacuate the area. This is why I am supporting the following discussion on life-saving rules and personal gas detectors.

Life-saving Rules and the Personal Gas Detector

I believe in these personal gas detectors/monitors and have referred to them throughout this book and in both of my previous Process Safety Books, *Process Operations Safety, The What, Why, and How, Behind Safe Petrochemical Plant Operations* and *The Hazards of Nitrogen and Other Inert Gases, and How They can be Safely Controlled*. There is little doubt that lives will be saved if these detectors are routinely checked to ensure full functionality and if they are always worn when personnel are in areas containing H_2S in high concentrations. This means these devices will be worn in all areas where the workers could be exposed to H_2S concentration at or above 10 ppm (the NIOSH PEL). These devices are typically set to alarm at between 5 and 10 ppm, and when the device alarms, everyone should immediately leave the area and not return unless they are wearing a self-contained breathing apparatus (SCBA).

Many facilities have what they refer to as "Life Saving Rules," "Critical Safety Rules," or other special designations, which are typically safety rules of the type that help guard against cases where one minor mistake in a procedure or one exposure, or other incidents of this type, can result in severe injury, a significant loss of containment, or even a loss of life. Generally, there are a relatively small number of critical rules that are mandatory and are considered a condition of employment. In other words, failure to comply with these rules can subject the employee to termination. I realize that this is stringent and may seem overly threatening; however, remember these rules are designed to protect the workers' lives.

I support the adoption of a policy at all facilities where high concentrations of H_2S either exist or can exist, requiring personnel to wear an H_2S detector anytime they are in areas that may have high concentrations of hydrogen sulfide. I believe this policy should be formally adopted by the facility and enforced as a Life-Saving Rule. This would mean that wearing the detector in a high H_2S area would be a condition of continued employment at the facility. It should be straightforward for most affected industries, such as oil and gas and municipal sewer plants, to enforce this through their existing internal organizational structure. I suggest that communication for the dairy and ranch operations could be done either through

Table 9.1 Examples of Life-Saving Rules (these are only examples and are not intended to be a complete list).

1. Always use approved fall protection when working at heights.
2. Obtain an approved hot work permit before performing hot work.
3. Always perform a gas test, have a standby attendant, and obtain proper authorization before entering a confined space.
4. Obtain authorization and ensure proper mitigation before disabling a safety-critical device.
5. Always verify proper equipment isolation before beginning work.
6. Always wear your seat belt when driving or riding in a vehicle.
7. No use of alcohol or drugs, and none in your system, while working or driving.
8. Absolutely no smoking, except in clearly marked designated areas.
9. Always wear a personal H_2S detector when in H_2S areas. **(PROPOSED)**

their existing management structure or through ranch and dairy cooperative associations.

In this book, I reviewed two incidents where the operator was assigned an H_2S detector but was not wearing it when they were exposed, and in both cases, it cost the operator their life; in one case, it also cost the life of the operator's wife. I believe that putting on an H_2S detector should be like putting on your seat belt; you shouldn't have to think about it. It should be an automatic action.

With this policy in place as a Life-Saving Rule, noncompliance can result in discipline, including dismissal. Again, I recognize that this is severe, but we are talking about saving lives. Unfortunately, we have experienced the deaths of workers who were provided with the detectors but were not wearing them when they were exposed. These are clear examples of how these devices will save lives when they are properly used.

Table 9.1 provides several examples of Life-Saving Rules. From my perspective, it should be straightforward to adopt a rule regarding the wearing of an H_2S detector as a Life-Saving Rule anytime the employee is inside the boundary of a process unit or operating area where a high concentration of H_2S may be present.

Verification of Hydrogen Sulfide Detector (Bump Testing)

Personal H_2S detectors must be periodically verified to ensure that they provide accurate and reliable readings. This is done by "bump testing" the device using a known reference gas, also called a calibration gas or "cal" gas, to verify that the sensors are reading the reference gas concentration and reporting the readings accurately. For H_2S detectors, the reference gas concentration is typically 25 ppm of H_2S in a mixture

Courtesy of Penn State Extension

Figure 9.5 Illustration of an H_2S bump test station and charging cradle.

of another benign gas, one that does not affect the H_2S concentration reading. The test is performed by affixing a short length of tubing to the sensor input from the reference gas. Once the test starts, the sensor and alarm should indicate the correct concentration within a specified time, typically less than one minute.

The facility should have a written procedure detailing the expected frequency and process for carrying out the bump tests for each detector. All bump tests should be documented in a log or similar tracking system located near the bump test station. Figure 9.5 illustrates an example of a bump test station with the bump test cradle and charger.

Please see the article by Penn State Extension, "Use Personal Gas Monitors to Avoid Exposure to Toxic Hydrogen Sulfide," via a link in the Additional Resources section.

Fixed Hydrogen Sulfide Detection/Alarms

Continuous fixed gas quality testing with visual and audible alarm capabilities should be considered in addition to personnel monitors, especially in areas subject to high concentrations of H_2S. These devices are designed with fixed H_2S sensors and alarms, including visual and audible alarms. All personnel working at the facility must be trained in the sound of the alarm and the actions to take in the event of a detected release, including immediate evacuation of the area by quickly evacuating the area and moving crosswind (never upwind or downwind).

The fixed alarms are available for permanent installation or temporary service. For example, I supported a construction project on the West Coast of the United States, which involved a very large and deep excavation for a new project foundation that was being built between, and very close to, other large and complex process units. This project was sandwiched between an HF Alkylation Unit, a Hydrogen Reformer, and a large Fluid Catalytic Cracking Unit. In this case, we contracted for continuous gas monitoring near the bottom of all four corners of the excavation.

Continuous gas testing was conducted for four of the major concerns of this work being close to these operating areas and substantially below grade.

To continually test to ensure continuous safe working conditions for the workers, we were continuously monitoring oxygen content (O_2), hydrogen sulfide (H_2S), hydrofluoric acid vapors (H-F), and, of course, the lower explosive limit (% LEL) for flammability. All tests were continually being run at all four corners of the excavation with alarms programmed for visual and audible indications if any trigger points were reached or exceeded by any of the monitored gases. This continuous testing was done for all periods when workers were inside the excavation and conducted at a very reasonable price for the company.

This clearly shows that this capability is readily available to industry and should be a routine part of most hazardous work areas where there is the potential for a lack of oxygen or the presence of toxic vapors.

Please see Figure 9.6 for a photograph of a fixed gas sensor. This example is a BW Technologies Rig Rat III Gas Detector (Non-Wireless Version). This is a multi-gas monitor that can be designed for H_2S or other toxic, flammable, or inert gases that may be at your facility. It has remote reading and provides audible and visual alarms.

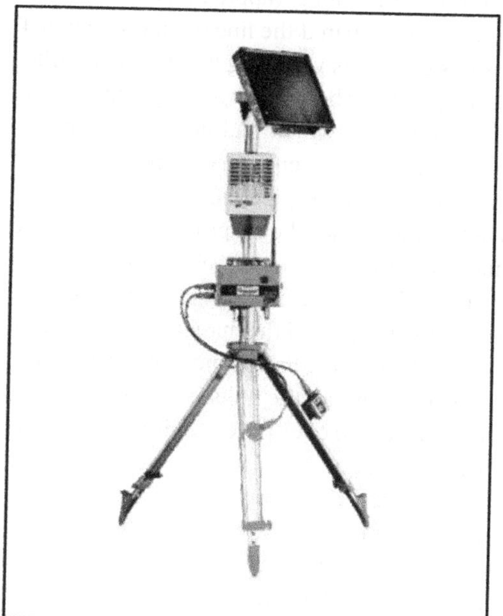

Courtesy of Honeywell and BW Technologies/JJ Technologies

Figure 9.6 A photograph of a BW Technologies Rig-Rat Gas Detector. (Please see links in the Additional Resources section.)

Requirement for Equipment Walk-down and Flange Identification

Anytime work is being considered involving opening process equipment on or near equipment that has contained or still contains a high concentration of hydrogen sulfide, a detailed walk-down must be completed between the work permit issuer and the people performing the work. Every flange to be opened must be identified and properly tagged with a distinctive tag identifying it as a point to be opened. The process I am familiar with uses a psychedelic-colored tag that is only used for this purpose, which we refer to as a "crazy tag." This tag was only hung by the operations group, which was responsible for preparing equipment for opening, and when we saw this tag, we understood that this was a flange that had been prepared and was ready for opening. This is important to help prevent the craftsmen from inadvertently getting off-track and opening a flange on an adjacent piping circuit that is pressurized and may contain high concentrations of hydrogen sulfide. The piping and other process equipment must be demonstrated to be at a zero-energy state, and vents and drains must be open and demonstrated to be unobstructed.

Although it happened a long time ago, an incident that occurred at our site occurred when a pipefitter accidentally, while removing blinds on a 4" (101.6 mm) propane pipeline, opened a flange on the very long (approximately 10,000 ft [3,048 m]) pipeline. The operations group had prepared the line on one side of a 4" valve for opening; however, the opposite side of this very long line remained filled and pressurized with liquid propane. The pipefitter began his work by removing all the bolts from the flange on the side of the valve, which had not been prepared. However, the flange remained stuck to the valve, and even with the line pressurized with liquid propane, it remained sealed. That is until the pipefitter opened the flange with a wedge and a hammer. As soon as the line opened, liquid propane was released and quickly vaporized, creating a massive flammable vapor release. Fortunately, there was no fire or explosion, but the vapor release affected a very large process area. Of course, with all the bolts removed from the flange and the massive release of propane, there was no way for the pipefitters to reclose the flange and stop the release.

This explains why workers should never remove all the bolts from a flange until the flange is cracked open and proven to be at zero energy. Removing all the bolts can leave the flange "stuck" in the closed position, and if it is pressurized, it can open suddenly and unexpectedly, resulting in a massive release when it opens. Cracking the flange open will allow the worker to quickly reclose the flange if it is found to be pressurized or if the product is leaking from the flange.

The investigation into this incident discovered that the work permit had been issued in the control room, and no walk-down had been carried out between the permit issuer and the people doing the work before the pipefitter opened the flange. The key learnings that resulted from this incident included ensuring that all flanges to be opened are properly tagged with a special tag indicating the line has been prepared for opening and that a detailed walk-down is completed on all equipment. Clear communication has been carried out between the permit issuer and the people doing the work to ensure they are familiar with the work and understand exactly which flanges are to be opened.

Another key learning is that craftsmen when opening a flange should always stand away from the line being opened and should never remove all the bolts until the flange has been loosened to ensure zero energy. This ensures that if pressure remains or liquid is discovered, the craftsmen can quickly reclose the flange and communicate this to the operations group. I also discussed that if the opening is a "first break" into the system, an operator or someone from operations should be present during the opening. These changes were made to permit forms, and training was completed to help ensure that this is done. Even though this incident happened a long time ago, these changes remain in place today. Hence, the crazy tag mentioned earlier. I discuss this incident in all my Safe Operations Training courses to ensure these key learnings are communicated to everyone who attends.

No hydrogen sulfide was released in this incident; however, the same thing could happen if the mechanics accidentally opened the wrong flange on a pipeline or other equipment containing high-concentration hydrogen sulfide. As stated earlier, several of these pipelines, especially those on or near the amine regenerators and sulfur plants, or from the regenerators to sulfur plants, contain hydrogen sulfide in concentrations exceeding 80–90% or even higher hydrogen sulfide. This means that concentrations in these pipelines may exceed 800,000–900,000 ppm. A release of this type can be catastrophic and most likely affect areas well outside the facility boundaries.

Anyone who is not protected from these highly toxic vapors with an SCBA can be killed instantly. This explains why these procedures are so important and must be implemented and followed by all sites with this kind of H_2S concentration in their process or interconnected pipeline systems. I am also aware that some refineries generating a high concentration of H_2S send super-concentrated H_2S to neighboring facilities for further processing, for example, to convert it to sulfuric acid. These pipelines have the same potential and should be handled in a similar fashion for all work on or near these systems.

At our site, the piping to and from the amine regenerators, and from the regenerators to the sulfur plants, all containing super-rich concentrations of hydrogen sulfide, was painted with distinctive colored bands around the piping. Any work on or near these pipelines, including other pipelines nested among the high-concentration H_2S lines, was deemed a high priority and received special safety considerations. This includes any planned lifting operations near these pipelines.

This is one more prevention step to stop someone from opening the wrong line. Suppose, for example, you can imagine how easy it is for workers to open flanges on a single piping system to install blinds and open the wrong pipeline.

After all, they are working in a major pipe rack with dozens of other pipelines; all are painted the same color, and most are the same pipe size and class, with no identification. The distinctive colored bands highlight these pipelines as special due to their super-rich concentrations of hydrogen sulfide.

Personnel Protection for Vessel Entry

Working in Enclosed Areas or Confined Spaces

It is extremely important to follow the OSHA confined space entry procedure, including all of the precautions: a gas test of the space before entry (oxygen content, flammables [LEL], hydrogen sulfide [H_2S]); an entry attendant must be present at the access point for the full duration that entry is underway; and a rescue plan must be developed, also before the entry. The gas test should be continuously monitored (preferred) or completed within two hours of entry. The OSHA confined space entry procedure can be quickly found online. For a link, please see the Additional Resources section of this chapter.

If the confined space is not designed for human entry and either contains or has the potential to contain a hazardous substance, like hydrogen sulfide, it qualifies as an OSHA permit-required confined space. This means that it must be gas-tested for percent oxygen, flammables (percent of LEL), hydrogen sulfide (ppm of H_2S), and other toxins specific to the space being entered.

This means that the atmosphere should be tested before entering into an excavation of more than 4 feet in depth. Also, remember that due to the nature of hydrogen sulfide, it should be expected in excavations or enclosures that contain, or has contained, animal waste, manure, any rotting or decaying vegetation, runoff from waste treatment plants or landfills, sour water, for example, from oil and gas wells, oil well condensate, or crude oils or naphtha from oil and gas wells. In addition to excavations, this also includes enclosures such as tank trucks, tank cars, storage tanks, grain silos, waste pits, and other similar enclosures that may contain rotting or decaying waste or vegetation. This is extremely important, and lives can be saved by taking these simple precautions. A quick review of Appendix 2 will exemplify the many lives lost by not following these procedures.

Entering Process Equipment (Towers, Drums, Pits, Excavations, Grain Silos, Tank Trucks or Tank Cars, Sewer or Electrical Manholes, Ship Cargo Holds)

Opening Process Equipment

Anytime personnel open process equipment that has been in high hydrogen sulfide service, the prescribed PPE should be based on current gas test results of the environment inside the piping or equipment. In most cases, this is expected to include the use of either hose-line supplied breathing air or an SCBA. Personnel exposure cannot exceed 10 ppm (the NIOSH PEL) in all cases unless they wear the appropriate PPE, such as SCBA.

Please remember that the gas test must be recent; a rule of thumb is that the gas test must either be continuous monitoring (preferred) or have been performed within two hours of the equipment opening. The test is always for oxygen first, as most gas test machines require a specified minimum amount of oxygen before they can provide a reliable reading for flammability. The gas test must also test for hydrogen sulfide, and if any equipment is to be entered, it must be tested for oxygen content, and all other precautions for confined space entry must be in place. This includes a standby attendant in direct contact with the entrant(s) who can summon rescue if needed, a personnel rescue plan and resources, and all other provisions of the OSHA regulation 29 CFR 1910.146.

It is also important that workers are provided with and wear appropriate PPE, such as chemical safety goggles, full-body protective suits, and an SCBA when warranted by the gas test. Each worker should also monitor air quality on an ongoing basis with a hydrogen sulfide detector.

A well-thought-out emergency response plan should also be established in the event of H_2S exposure. Forced air ventilation should be used wherever possible to help evacuate the toxic vapor away from the workers.

The OSHA regulatory requirement for the PEL is 20 ppm (OSHA PEL) (10 PPM NIOSH REL) and should not be exceeded at any time. Also, 50 ppm is the maximum peak and should not be exceeded during any 10-minute work period. Also, NIOSH (National Institute of Safety and Health) has established the recommended airborne exposure limit (REL) as 10 ppm, which should not be exceeded during any 10-minute work period. Several facilities that I am familiar with have established internal standards below these limits and you should be familiar with those requirements at your site.

Most of these (1) are large enough for a worker to enter, (2) have limited means of entry or exit, and (3) are not designed for continuous occupancy. Add the potential for hazardous materials to be present, and you have the definition for a permit-required confined space. Before attempting to enter a confined space, you must ensure that all the requirements for a permit-required confined space entry have been met.

Anytime personnel enter a confined space that has contained or contains a flammable or hydrogen sulfide, forced air ventilation is extremely important. This is typically done using a forced air system consisting of a fan or blower equipped with flexible air ducts to ensure the forced air is properly routed to the entrants. If this system is used in process areas where there is a possibility of flammable vapor being present or in electrically classified areas, the motor should be consistent with an electrical Class 1, Division 1 classification.

Most importantly, there must be a gas test of the environment inside the confined space to ensure that it will support life. This means a gas test for oxygen content (result must be 19.5% or more, but less than 23% oxygen), for flammable (must be less than 10% of the LEL), and for hydrogen sulfide (must be less than 20 ppm [OSHA PEL] or 10 ppm [NIOSH H_2S REL]). The 20-ppm maximum for hydrogen sulfide from the OSHA gas tests for other toxic materials specific to the process must also be made before vessel entry.

There must be a standby attendant who is in direct contact with the entrant(s) throughout the confined space entry. The attendant must be prepared to summon rescue in the event the entrant(s) require it.

There must be a personnel rescue plan in the event an entrant needs assistance exiting the confined space or in the event of another emergency. The rescue plan should be documented, and rescue personnel, materials, and equipment should be readily available at the job site.

Please see Figure 9.7 for an example of a confined space entry that went tragically wrong. This incident involved exposure to nitrogen; images of exposures to hydrogen sulfide are rare; however, the same thing can happen with hydrogen sulfide (H_2S).

This is an example of a confined space being purged with nitrogen. Two workers entered this section of pipe to perform an internal inspection using a black light. The end of the piping was covered with plastic sheeting and a black tarp to keep the light out. There was no gas test, no confined space entry permit, and no standby person. One worker was killed, and one was seriously injured but survived.

US Chemical Safety and Hazard Investigation Board (CSB)/Public Domain

Figure 9.7 Union Carbide Corporation, Hahnville, Louisiana. March 27, 1998. Photograph of a section of open-ended piping undergoing a nitrogen purge used to protect the catalyst from moisture. One worker was killed, and one was severely injured due to exposure to nitrogen (lack of oxygen). At the time of the incident, the end of the pipe was covered with a black tarp to keep the light out.

Personnel Rescue

Anytime we work with teammates around or near equipment containing high-concentration hydrogen sulfide, we should be very aware of what our workmates are doing and their actions. If we observe one of our teammates collapse and fall to the ground or the deck, we must resist the human urge to rush in and provide aid or assistance. Instead, we must fall back on our training. We must recognize that there is a high potential that we have witnessed an exposure to high concentration of hydrogen sulfide and that the person has succumbed to the toxic nature of the deadly gas. Unless we are properly trained in rescue techniques and are equipped with a personal SCBA, we should NEVER rush in to save our coworker. All the data on this clearly indicates that the likely outcome is that we will also become a victim, possibly collapsing on top of the coworker and delaying their rescue and resuscitation. There are several examples of this in Appendix 2, where the would-be rescuer also died, and occasionally where the second, third, and even fourth rescuer also died due to the detected presence of hydrogen sulfide.

Unless we are trained in rescue techniques and equipped with proper protective equipment such as an SCBA, instead of rushing in to provide assistance, we should immediately call for emergency response and then stand by to direct the rescue personnel to the victim. Stay clear of the area until the responders arrive. They will arrive fully outfitted in emergency breathing air and with the proper rescue tools and equipment.

Self-contained Breathing Apparatus

I have mentioned the use of SCBA quite a bit thus far in this book and I felt I should explain in more detail the importance and the role they can play in saving lives. In many of the case studies and examples where fatalities occurred and where hydrogen sulfide played a role, you heard me say that "only if they had been wearing an SCBA they would have survived, and deaths would not have occurred." And, of course, this is true; SCBAs have and can save lives if they are used properly.

I discussed this briefly in my second Process Safety Book, *The Hazards of Nitrogen and Other Inert Gases and How They Can Be Safely Managed*. I apologize to readers who have already read this book, but this topic is so important that I could not leave it out here.

If personnel are expected to work near vessel openings or open pipe flanges where hydrogen sulfide may be present or where nitrogen or other inert gases are venting, consideration should be given to requiring an SCBA for those personnel, with a standby person also in SCBA. See Figure 9.8 for an example of an SCBA.

As mentioned earlier, any attempted rescue of a person who has collapsed on a process unit or process vessel, where there is a possibility of hydrogen sulfide, another toxic gas, or nitrogen or another inert gas being present, should always

Courtesy of MSA Safety

Figure 9.8 Example of an SCBA with a full-face mask.

respond wearing an SCBA. The best-case scenario is they should be wearing an SCBA, and they have a backup person who is also wearing an SCBA when responding in a rescue situation.

SCBAs are typically available in negative-pressure or positive-pressure configurations and several sizes. Generally, the higher the tank pressure, the longer the response or rescue breathing time.

I prefer the positive-pressure regulator in combination with a full-face mask. The positive pressure will supply air at pressure to the mask anytime the pressure difference is reduced. The pressure inside the mask is slightly above ambient pressure, ensuring positive airflow is always available to the user. This can overcome minor leaks in the mask, preventing the leakage of smoke or vapors into the mask.

The negative-pressure system is quickly becoming obsolete; however, it may still be available in some areas. The disadvantage of the negative-pressure regulator is that the user must activate the regulator by creating negative pressure when breathing. In some situations, this may allow outside smoke or toxic vapors into the mask if the mask is leaking.

Hydrogen Sulfide Flammability and Hot Work

By now, I know you recognize that the number one hazard with hydrogen sulfide is its toxicity and that one breath of this deadly gas in high concentrations can, at a minimum, knock a person down and that it can kill. While this is the number one hazard, hydrogen sulfide is also a very flammable gas. It has an extremely low flash point and a very wide flammable range, making it extremely dangerous from a flammability perspective. The flash point is −116 °F (−82 °C), and the flammable range is 4.3–45.5%, making it an extremely dangerous gas due to its easily ignitable characteristics. People have been severely burned and killed by being caught up in a fire fueled by hydrogen sulfide. There are a couple of case studies in Appendix 2 of exactly this happening, where workers died due to burn injuries.

Anytime hot work is planned in areas where hydrogen sulfide may be present, a gas test must be conducted for both hydrogen sulfide and flammables (LEL). No confidence should be placed in your ability to smell the distinctive odor of rotten eggs due to the potential for the odor senses to be affected by the H_2S.

Additional Resources (How H_2S Hazards Are Controlled)

Occupational Safety and Health Administration (OSHA), 1910.146 – Permit-required confined spaces, https://www.osha.gov/laws-regs/regulations/standardnumber/1910/1910.146

Modular designed multi-gas detection system, BW Technologies Rig Rat III Gas Detector (Non-Wireless Version), BW Technologies Rig Rat III Gas Detector (Non-Wireless Version), ttps://www.jjstech.com/rr-3000s.html?gad_source=1&gclid=Cj0KCQiA4fi7BhC5ARIsAEV1YiYwzcGqFW0KaWq8ceyjgGMcctHRbCPI4hM3ThD8gdmv2zhlsWZ-yQUaAgXDEALw_wcB

Forensics Detectors, https://www.forensicsdetectors.com/?gad_source=1&gclid=EAIaIQobChMIxqb87KSWigMVNSFECB05kw44EAAYASAAEgJ0IPD_BwE

MSA The Safety Company, MSA Self-Contained Breathing Apparatus, https://us.msasafety.com/g1scba-demo?utm_source=google&utm_medium=cpc&utm_campaign=g1-scba&utm_content=fs&gad_source=1&gclid=EAIaIQobChMIvYPJqYOOigMVFQ5ECB3kiifeEAAYASAAEgLBhfD_BwE

US Occupational Safety and Health Administration (OSHA), Hydrogen Sulfide; Evaluating and Controlling Exposure, https://www.osha.gov/hydrogen-sulfide/evaluating-controlling-exposure

US Chemical Safety and Hazard Investigation Board, Union Carbide Corp. Nitrogen Asphyxiation Incident, https://www.csb.gov/union-carbide-corp-nitrogen-asphyxiation-incident/

Penn State Extension, Use Personal Gas Monitors to Avoid Exposure to Toxic Hydrogen Sulfide, https://extension.psu.edu/use-personal-gas-monitors-to-avoid-exposure-to-toxic-hydrogen-sulfide

Penn State Extension, Manure Gas Safety, https://extension.psu.edu/manure-gas-safety

Jason Hessom (Published on July 21, 2022), 13+ Best H_2S Safety Tips for 2023 – [Hydrogen Sulfide], https://www.safetyvideos.com/h2s-safety-tips#:~:text=Use%20Personal%20Protective%20Equipment,workers%20from%20hydrogen%20sulfide%20exposure

Chapter 9: End-of-Chapter Quiz

1. What is the purpose of the jobsite walkdown prior to opening process equipment?

2. When would it be acceptable to attempt to rescue or provide aid to a stricken coworker, one who collapsed while working on the jobsite?

3. What are the three key elements of a successful confined space entry program?

4. What is the proper response if a coworker collapses while on the jobsite?

5. What does a successful gas test for oxygen look like?

6. What is the ideal approach for evacuating vapors from the worksite when opening process equipment?

7. What type of personal protective equipment should the workers use when opening process equipment?

8. The purpose of the standby attendant during a confined space entry is to remain in direct ____________ with the entrant(s) throughout the confined space entry. The attendant must be prepared and ready to summon for ____________ in an event this is required by the entrant(s).

9. A detailed workplan incorporating jobsite safety when working around or on systems and equipment containing, or that has contained, high concentrations of hydrogen sulfide can save __________ and prevent tragedy that affects families for a __________, and beyond.

10. A good place to start addressing our long history of hydrogen sulfide poisoning incidents is with __________ and __________.

11. Initial hydrogen sulfide awareness __________ should be provided for all personnel who may be required to work in areas where H_2S is processed or handled, including an annual ____________ training.

12. Everyone who works at a facility where hydrogens sulfide is produced, stored, or handled in high concentrations should be equipped with a personal H_2S __________ that has both __________ and __________ alarm capability.

13. Prior to entry into an excavation more than ______ feet in depth, the atmosphere should be tested ______ to entry.

14. To ensure success, the requirement to wear a personal H_2S detector should be formally adopted by the facility and enforced as a safety rule that wearing the detector is a condition of continued ______ at the facility.

15. Facilities that have processing equipment with high concentrations of H_2S should be equipped with fixed H_2S sensors with ______ and ______ alarms in the event of an H_2S release.

16. All personnel working at the facility must be trained in the alarm sound and ______ to take in the event of a detected release, including immediate evacuation of the area by moving crosswind (never ______ or ______).

17. The flash point for hydrogen sulfide is ______, and the flammable range is ______% to ______%, making it extremely ______.

10

Recognition of International and Regional Authorities/Organizations with Expertise in Hydrogen Sulfide

Authorities/Organizations with Expertise in Hydrogen Sulfide (Not Intended to Be a Complete List)

- Energy Safety Canada (https://www.energysafetycanada.com/)
- The US Chemical Safety and Hazard Investigation Board (CSB) (https://www.csb.gov/)
- The UK Health and Safety Executive (HSE) (https://www.hse.gov.uk/)
- The Occupational Safety and Health Administration (OSHA) (https://www.osha.gov/)
- The National Institute of Safety and Health (NIOSH) (https://www.cdc.gov/niosh/index.html)
- American Conference of Governmental Industrial Hygienists (ACGIH) (https://www.acgih.org/)
- American Industrial Hygiene Association (AIHA) (https://www.aiha.org/)

Hydrogen Sulfide (H_2S) Hazards: Risk Recognition and Prevention, First Edition. M. Darryl Yoes.

Epilogue

New York Harbor/Near West 138th Street

Sewage Barge Explosion

New York City Department of Environmental Protection (DEP)

One fatality and two injuries, including one treated at a local hospital

May 24, 2025

After I finished the manuscript for this book, I sent it to the publishers so they could do their magic when another tragic incident possibly involving hydrogen sulfide occurred in the harbor of New York City. As I have said several times during the book, incidents like these occur regularly but rarely at the same location or as repeated incidents. In most of these incidents, we may not be aware of what has happened or may not have enough information to connect the dots and understand the role that hydrogen sulfide plays in what has happened. This may be what happened in this most recent incident, which occurred this past weekend in New York City. In the public reports, there was no indication of hydrogen sulfide being involved in this incident; however, when you hear what happened, I believe you will agree that hydrogen sulfide may be involved.

On May 24, 2025, a 59-year-old New York City Department of Environmental Protection (DEP) worker was killed in a barge explosion on the Hudson River near West 138th Street while transporting raw sewerage. According to local media, two others sustained injuries in the explosion, with one being treated for injuries at a local hospital. The explosion happened when the barge carried what was reported to be raw sewage from New York City to an offsite treatment facility and docked near 138th Street for maintenance repairs. Local media also reported that welding

Hydrogen Sulfide (H_2S) Hazards: Risk Recognition and Prevention, First Edition. M. Darryl Yoes.

operations were in progress on or near the barge, which may have been the ignition source, although this has not been confirmed.

Raw sewage and other sewage waste generate methane gas, which is extremely flammable. If it is raw sewage, it also generates copious amounts of hydrogen sulfide, which is also extremely flammable. Both methane and hydrogen sulfide are explosive when they are in the flammable range – the correct ratio of fuel to air – sometimes also called the explosive range. Both methane and hydrogen sulfide have been responsible for catastrophic explosions in other incidents. An investigation is underway for this incident, but it may be too early, and there is not enough data to confirm the cause.

The public reports indicate there was hot work and possibly welding in the area and possibly on the barge. This and any other potential ignition sources should be investigated thoroughly. Any work involving ignition sources should be avoided near or on the barge due to the known methane and H_2S flammability. Before hot work is considered, it should involve detailed planning and precautions to prevent the ignition of flammable vapors. This should include gas testing and a hot work permit to authorize the hot work.

The investigation of this incident should be thorough to understand what happened, but most importantly, to help prevent another similar incident from happening. This should include reviewing the documents at the scene, including the hot work permit and gas test results if they exist, and interviewing survivors to better understand what they were doing and how this incident occurred. Even though the worker was experienced, the investigation should review and confirm the worker's training and qualifications. The investigation should also include recommendations or corrective actions that address the findings and conclusions.

When the investigation is completed, the results should be shared broadly. Unfortunately, this is rarely done unless the US Chemical Safety Board or other federal agencies are involved.

Regardless of whether this turns out to be methane, hydrogen sulfide, or another flammable material, the questions that should be answered in this investigation include the following:

- Was the barge carrying raw sewage, as reported by the media, or was it something else?
- If the barge was carrying raw sewage, was the potential generation of hydrogen sulfide considered?

- Was welding or other hot work in progress near or on the barge carrying sewage?
- If welding or other hot work was not in progress, what other potential ignition sources were present in the area?
- Were considerations made for hot work permits for welding or other hot work in this area?
- If welding or other hot work was in progress, was it properly authorized with a hot work permit?
- Was a gas test completed before authorizing welding or other hot work near the sewage barge before the explosion?
- If a gas test was conducted before the explosion, did it include testing for hydrocarbons such as methane and hydrogen sulfide (H_2S)?
- Are gas testing results available for prior shipments of sewage from the same location or source, and do they confirm the presence of methane or hydrogen sulfide?

Appendix 1

End-of-Chapter Quiz (With Answers)

Chapter 1: End-of-Chapter Quiz (With Answers)

1. Where are we likely to find significant amounts of hydrogen sulfide? Please choose all that apply.
   a. Swamps or bogs with rotting vegetation
   b. Oil and gas well condensate
   c. Liquids draining from waste disposal sites
   d. Sour water drained from crude oil storage tanks
   e. Agricultural grain silos with rotting corn or grain
   f. None of the above
   g. All of the above
   Answer: g. All of the above
2. In which industries are we most likely to find significant concentrations of hydrogen sulfide?
   Please choose all that apply.
   a. Leather tanning
   b. Petroleum refining
   c. Wastewater treatment
   d. Pulp and paper mills
   e. Textile manufacturing
   f. Mining
   g. None of the above
   h. All of the above
   Answer: h. All of the above
3. When a sour water stream, water containing hydrogen sulfide, contacts ______________, a significant amount of hydrogen sulfide gas may be released into the atmosphere.
   Answer: acid

Hydrogen Sulfide (H_2S) Hazards: Risk Recognition and Prevention, First Edition. M. Darryl Yoes.

4. Most petroleum refineries process hydrogen sulfide in concentrations as high as ________ to ________ percent.
   Answer(s): 80, 90 (800,000–900,000 ppm).
5. Major H_2S releases have occurred by the accidental ____________ of a truckload of incompatible chemicals into the wrong tank.
   Answer: unloading
6. The most common way H_2S is removed from the product in a petroleum refinery is by scrubbing the product with an ________ chemical.
   Answer: amine
7. Please describe the purpose and function of a hydrotreater unit in a petroleum refinery.
   Please select the most correct answer.
   a. The hydrotreater removes sulfur compounds from the crude oil before it goes to the crude units.
   b. The hydrotreater ensures that the product meets consumer specifications.
   c. The hydrotreater removes sulfur compounds from the naphtha product using fixed catalyst, hydrogen, and heat.
   d. The hydrotreater is the most important unit because it produces clean hydrogen.
   Answer: c. The hydrotreater removes sulfur compounds from the naphtha product using fixed catalyst, hydrogen, and heat.
8. What function does the amine absorber play in a petroleum refinery hydrotreater unit?
   Please select the most correct answer.
   a. The amine absorber sends the foul amine to the sulfur plant.
   b. The amine absorber removes the hydrogen sulfide from the gas streams by absorbing it onto amine.
   c. The amine absorber cleans the foul amine.
   d. The amine absorber is in the feed system to the hydrotreater and purifies the feed to the reactor.
   Answer: b. The amine absorber removes the hydrogen sulfide from the gas streams by absorbing it onto the amine.
9. What is likely to happen if sour water (water containing H_2S) is mixed with any form of acid?
   Answer: When sour water is mixed with acid, significant amounts of hydrogen sulfide will be released.
10. Hydrogen sulfide is also found in ________ treatment plants and underground sewer and wastewater piping and pump transfer operations. Also, large ____________ ranches and ________ farms may accumulate animal waste and temporary pits or storage tanks, and typically the waste will

generate significant accumulations of hydrogen sulfide in the vapor space ________ the accumulated waste.
Answer(s): wastewater, cattle, dairy, above

11. More typical in summer, the hot weather significantly ________ the decay of most organic materials, such as manure and vegetation. This ________ the hydrogen sulfide generation.
Answer(s): accelerates, increases

Chapter 2: End-of-Chapter Quiz (With Answers)

1. At wastewater treatment facilities, hydrogen sulfide can be produced from the ________ breakdown of organic matter in the absence of ________.
Answer(s): microbial, oxygen
2. At wastewater treatment plants, the breakdown of organic matter is referred to as ________ ________ and occurs in equipment referred to as anaerobic digesters.
Answer(s): anaerobic digestion
3. It is also extremely important for dairy farmers, cattlemen, and ranchers to understand that hydrogen sulfide (H_2S) can be readily formed from the natural decay processes, or the ________ process in cattle or horse manure.
Answer: biodegradation
4. This occurs when bacteria, especially ________-________ (SRB), break down organic matter containing sulfur compounds, such as proteins.
Answer(s): sulfate-reducing bacteria
5. The sulfate-reducing bacteria (SRB) can convert sulfate in the wastewater to sulfide, which then combines with ________ to produce hydrogen sulfide gas (H_2S).
Answer: hydrogen
6. Hydrogen sulfide is primarily formed when the biodegradation occurs in an ________ environment. This means that most H_2S production due to biodegradation happens in environments with little or no ________.
Answer(s): anerobic, oxygen
7. ________s are known to play a crucial role in the process by converting sulfate into sulfide, which then reacts with hydrogen to form H_2S.
Answer: SRB
8. After the bacteria break down the organic matter, the largest portion is primarily sand, loam, and clay, while the organic fraction is known as ________.
Answer: humus

9. Significant amounts of hydrogen sulfide can be released when the decaying matter is stirred or __________ and can result in the __________ of workers or others who may be nearby.
   Answer(s): disturbed, deaths
10. Hydrogen sulfide can also occur deep within landfills where there is a lack of __________.
   Answer: oxygen

Chapter 3: End-of-Chapter Quiz (With Answers)

1. Hydrogen sulfide (H_2S) is typically a __________ and __________ gas that in low concentrations has a very distinct __________ __________ odor and in low concentrations it is very noticeable due to this objectionable __________.
   Answer(s): colorless, toxic, rotten egg, odor
2. What is significant about hydrogen sulfide concentration of 0.01–1.5 ppm? Please select the most correct answer from the following.
   a. This is the concentration when the odor is noticeable to most people.
   b. There is really nothing significant about this concentration of H_2S.
   c. 0.01–1.5 ppm is the concentration when the olfactory nerves are paralyzed, and the odor is no longer noticeable.
   d. At this concentration, it is time to evacuate all personnel from the immediate area.

   Answer: a. This is the concentration when the odor is noticeable to most people.
3. At concentrations between __________ and __________ ppm, it paralyzes the __________ nerves, and the victim can no longer smell it.
   Answer(s): 100, 150, olfactory
4. Hydrogen sulfide is __________ than air and will settle in __________ places.
   Answer(s): heavier, low
5. Concentrations between __________ and __________ ppm result in almost instant death.
   Answer(s): 1,000, 2,000
6. Hydrogen sulfide is toxic, but it is also flammable, and in certain concentrations it can be __________.
   Answer: explosive
7. Hydrogen sulfide is not only __________, but it has a wide __________ __________, meaning that it is much easier to ignite.
   Answer(s): flammable, flammable range

8. Hydrogen sulfide is easily ignited by a nearby ignition source such as ________, a passing ________, or static ________.
   Answer(s): welding, vehicle, electricity
9. When present at concentrations above about 100–150 ppm, the only way a victim will know that this deadly gas is present is with a personal hydrogen sulfide gas ________ or a fixed ________ ________ monitor with alarm.
   Answer(s): detector, hydrogen sulfide
10. We all have an ingrained instinct that when we see a work mate or associate collapse, we have the human urge to ________.
    Answer: rescue
11. The flammable range for hydrogen sulfide is ________% LEL to ________% UEL.
    Please select the correct ratio of fuel to air for hydrogen sulfide.
    a. 6.2% LEL to 21% UEL
    b. 4.3% LEL to 45.5% UEL
    c. 2.5% LEL to 51.7% UEL
    d. 4.4% LEL to 11.5% UEL
    Answer: b. 4.3% LEL to 45.5% UEL
12. Where can we find the documented characteristics of hydrogen sulfide?
    Answer: The Safety Data Sheet
13. Hydrogen sulfide has a very ________ flash point and is easily ignited.
    Answer: low

Chapter 4: End-of-Chapter Quiz (With Answers)

1. Hydrogen sulfide becomes a serious risk to workers when it occurs or is ________ in confined spaces or areas with ________ or ________ ________ circulation.
   Answer(s): concentrated, little, no air
2. The common mode of exposure for humans is by ________ of the vapors. The H_2S gas is rapidly absorbed by the ________.
   Answer(s): inhalation, lungs
3. The victim may assume since they can no longer smell the H_2S that it has ________ or has gone away.
   Answer: dissipated
4. Asphyxiation by H_2S can happen in excavations or pits, especially in the following.
   Examples of areas where asphyxiation by H_2S can occur include the following:
   - Sewers or sewer treatment plants
   - Manholes

- Septic tanks
- Manure pits
- Municipal wastewater treatment plants
- Waste disposal sites
- Anywhere vegetation or other organic matter may be biodegrading
- Bogs or other similar areas with little or no oxygen
- Around or in oil and gas wells or petroleum refineries
- Crude oil tank farms
- Natural formations in the ground, which by their nature may be confined spaces

5. Why is hydrogen sulfide so deadly?
   Answers may include the following – not intended to be a complete list.
   - Hydrogen sulfide (H_2S) is an extremely dangerous and elusive gas!
   - It is highly toxic, flammable, and explosive.
   - Exposure to high concentrations can result in instantaneous unconsciousness and death to victims.
   - Exposure to concentrations as low as 100–150 ppm can result in paralysis of the olfactory nerves, which deadens the sense of smell where the victims will no longer realize that they are exposed.
   - It is a flammable vapor and explosive and has a wide flammable range, making it easily ignited. Concentrations of H_2S have exploded when in the flammable range (ratio) of H_2S and air.
   - It is toxic, even at low concentrations.
   - It is heavier than air and will travel along the ground and can accumulate and concentrate in low areas such as pits, sewers, underground vaults, etc.
   - It will quickly completely disable both victims and rescuers. Many rescuers have been killed while attempting to rescue a downed workmate.
6. According to the Bureau of Labor Statistics (BLS), hydrogen sulfide is a primary cause of ________________ in the United States.
   Answer: asphyxiation
7. Petroleum refineries typically have foul amine and hydrogen sulfide circulating in piping systems at concentrations in the 80–90% range or 800,000–900,000 ppm. One breath of H_2S at this concentration results in ________________ death.
   Answer: instantaneous
8. Hydrogen sulfide occurs naturally in sewers or manure pits, around or in oil and gas wells, and in natural formations in the ground, which by their nature may be __________ __________, these are particularly __________ for workers.
   Answer(s): confined spaces, hazardous

9. Sanitation workers have been exposed to and have died while cleaning or working in or near sewers, septic tanks, or municipal wastewater treatment plants due to the unexpected presence of ____________ ____________.
   Answer: hydrogen sulfide
10. What is the flammable range for hydrogen sulfide?
    Answer: 4.3% LEL to 45.5% UEL

Chapter 5: End-of-Chapter Quiz (With Answers)

1. Which US federal agency established the Immediately Dangerous to Life and Health limit for hydrogen sulfide?
   Choose the most appropriate response from the following.
   a. The US Occupational Safety and Health Administration (OSHA)
   b. The US Environmental Protection Agency (EPA)
   c. The US National Institute for Occupational Safety and Health (NIOSH)
   d. The US Centers for Disease Control and Prevention (CDC)
   e. The US National Institute of Health (NIH)
   Answer: c. The US National Institute for Occupational Safety and Health (NIOSH)
2. The hydrogen sulfide permissible exposure limit is established by US OSHA. Is it enforceable?
   a. Yes
   b. No
   Answer: Yes, it is enforceable under penalty law.
3. Please list the enforceable exposure limits (PEL) established by the US OSHA for hydrogen sulfide in industry.
   Answers(s): OSHA PELs:
   - General Industry Ceiling Limit: 20 ppm
   - General Industry Peak Limit: 50 ppm (up to 10 minutes if no other exposure during shift)
   - Construction 8-hour Limit: 10 ppm
   - Shipyard 8-hour Limit: 10 ppm
4. The US Occupational Safety and Health Administration (OSHA) reports that "Hydrogen sulfide gas causes a wide range of health effects. Workers are primarily exposed to hydrogen sulfide by ____________ it (by ____________ of the toxic vapors)."
   Answer(s): breathing, inhalation
5. The effects depend on how ________ hydrogen sulfide the victim breathes and for ______ _______."
   Answer(s): much, how long

6. OSHA also states "exposure to very _______ concentrations (of hydrogen sulfide) can quickly lead to _________."
   Answer(s): high, death
7. What is the Immediately Dangerous to Life and Health limit for hydrogen sulfide?
   Answer: 100 ppm
8. What is the eight-hour working limit for H_2S set by OSHA for
   a. general industry?
   b. shipyards?
   c. construction?
   Answer(s):
   a. 20 ppm
   b. 10 ppm
   c. 10 ppm
9. The peak H_2S exposure limit set by OSHA for general industry is _______ ppm for a maximum period of _______ minutes, provided there are _______ other exposure during the shift.
   Answer(s): 50 ppm, 10 minutes, no
10. Some industrial facilities abide by the NIOSH-recommended exposure limit of _______ ppm. Some sites have voluntarily reduced their internal standards to less than _______ ppm.
   Answer(s): 10 ppm, 10 ppm

Chapter 6: End-of-Chapter Quiz (With Answers)

1. When you think about the hazards of hydrogen sulfide and its toxicity, H_2S is primarily an _________ hazard.
   Please select the most correct answer:
   a. toxic
   b. skin and eye irritation
   c. inhalation
   d. neurological
   Answer: c. inhalation
2. A release of high concentration of hydrogen sulfide, generally 1,000 ppm or higher, can kill _________. In these concentrations, the victim is knocked down with the first _________ and they must be revived within ____ to ____ minutes.
   Answer(s): instantaneously, breath, 2, 4

3. All personnel, when in the field where _________ _________ is being processed or is present, even in moderate concentrations (concentrations of 10 ppm or more), should wear a H_2S _________ that has been bump-tested against a known _________ to ensure that it is fully functional.
   Answer(s): hydrogen sulfide, detector, standard
4. When personnel are in the process of opening process equipment for insertion of blinds, blind removal, opening access plates, or any other act of OPE (opening process equipment), the vessel or piping system must be verified that there is no _________ or substantial _________ present (_________ energy).
   Answer(s): pressure, product, zero
5. Each point of opening for blind insertion, blind removal, access plate removal, or any other vessel or piping system opening must be clearly identified by the _________ or operations organization. Generally, this is done by hanging a _________ or _________ tag with a _________ design and only used for this purpose on the flange to be opened.
   Answer(s): process, wearproof, weatherproof, unique
6. When sampling streams containing H_2S, personnel must wear proper personal protective equipment (PPE), including the following
   Your response should include the following – not intended to be a complete list:
   - Impervious gloves
   - Goggles (not required if wearing a full-face respirator)
   - Chemical-resistant clothing
   - Personnel should be wearing a calibrated and bump tested H_2S detector with continuous monitoring capability and audible, visual, and vibration alarms.
   - The breathing zone must be protected by wearing a full-face, positive-pressure, self-contained breathing apparatus (SCBA). The positive pressure prevents hydrogen sulfide leaks into the face mask.
   - It is highly recommended that the person collecting the sample be outfitted in a positive-pressure SCBA and a standby person also wearing a SCBA is located out of the immediate area, but who has a clear view of the operation.
7. What is one of the most important considerations when responding to a release potentially containing hydrogen sulfide?
   Please choose the most important consideration:
   a. The atmospheric temperature and humidity
   b. The wind direction
   c. The process pressure at the time of the release
   d. Ensuring that we can keep the unit in service
   Answer: b. The wind direction. Never attempt to approach a release from downwind.

8. When working around high concentrations of hydrogen sulfide or if sampling high concentrations of hydrogen sulfide, always have an emergency plan and ready availability to medical assistance if needed.
   a. For example, if an unexpected release occurs, immediately move all personnel ________ of the release and immediately notify the ________ ________ ________.

   Answer(s): upwind, emergency response team
9. When responding to a release, never venture into the area until you have properly donned a ____-________ ________ ________. With this done, you can determine what is being released, confirm if the release is flammable or toxic, and start development of a response plan.
   Answer(s): self-contained breathing apparatus (SCBA)
10. What is the primary benefit of using an engineered closed-loop sample system when sampling toxic products, for example, products containing high concentrations of hydrogen sulfide?
    a. An engineered sample system ensures that the sample is representative of the process.
    b. This type of system saves costs as the sample containers are reusable.
    c. Product losses are minimized with an engineered system.
    d. The operator is out of harm's way when the sample is being caught.
    Answer: d. The operator is out of harm's way when the sample is being caught.

Chapter 7: End-of-Chapter Quiz (With Answers)

1. Please name at least four process safety systems and facilities that have been significantly improved over the past twenty-plus years and should have reduced the number of hydrogen sulfide fatalities over this same period.
   Answers should include four of the following; not intended to be a complete list:
   - Process Safety Management System (PSMS)
   - Permit to Work Standard (PTW)
   - Confined Space Entry Standard
   - Operating Procedures
   - Process Hazard Analysis (PHA)
   - Management of Change (MOC)
   - Control of Defeat of Safety Critical Devices
   - Prestart up Safety Review (PSSR)
2. Please provide one example that indicates that lives continue to be lost due to loss of containment of the highly toxic hydrogen sulfide gas into the workplace.

Answer(s): I realize that this is a tough question due to the way it is worded. Answers like the following will get to the heart of this question.

- The human urge to rescue a teammate when he/she collapses.
- The October 2024 Pemex release where two workers died.
- Any other recent release resulting in a fatality or severe injury.

3. About how high can the hydrogen sulfide concentration be in a typical petroleum refinery hydrogen sulfide treating system?

   Answer: Typically, around 80–90% hydrogen sulfide, which is equivalent to 800,000–900,000 ppm. This occurs with equipment such as the amine regenerators.
4. How does 80–90% hydrogen sulfide convert to ppm?

   Answer: 800,000–900,000 ppm
5. What is the potential outcome if two incompatible chemicals are inadvertently mixed, for example, unloading the wrong chemical into a chemical storage tank?

   Answer: Mixing two or more incompatible chemicals can result in a severe release of toxic gases such as hydrogen sulfide or chlorine.
6. What atmospheric gas testing should be completed before entry into any process vessel, storage tank, tank truck or tank car, or any excavation 4 feet or more in depth?

   Answer: Atmospheric gas testing for any confined space prior to entry should always include oxygen (O_2), lower explosive limit (LEL), hydrogen sulfide (H_2S), and any other chemical or toxic that may be part of the process or the vessel's history.
7. What is the definition for an OSHA permit-required confined space? Provide definition as per OSHA 29CFR 1910.146:

   Answer: Permit-required confined space (permit space) means a confined space that has one or more of the following characteristics:
   1. Contains or has a potential to contain a hazardous atmosphere;
   2. Contains a material that has the potential for engulfing an entrant;
   3. Has an internal configuration such that an entrant could be trapped or asphyxiated by inwardly converging walls or by a floor that slopes downward and tapers to a smaller cross-section; or
   4. Contains any other recognized serious safety or health hazard.
8. List six examples of enclosures that meet the OSHA definition of a permit-required confined space.

   Answers should include examples similar to the following (not intended to be a complete list):
   - Storage tank
   - Railroad tank car

- Tank truck
- Process drum or other process vessel
- Large diameter piping

9. Approximately 15–20% of all fatalities involving hydrogen sulfide result in more than one fatality; occasionally, it involves as many as three or four fatalities. Why is this the case?
   Answer: We all possess a human urge to attempt rescue when we see a teammate or workmate collapse and become unconscious. When the person who is being rescued collapses due to hydrogen sulfide, the rescuer also becomes a victim and is frequently also killed.
10. When a truck driver is overcome by H_2S while sampling or gauging a tank truck, they are frequently killed by falling from the truck. Is the cause of an incident like this the failure to recognize the H_2S hazard, or is it failure to wear fall protection?
    Answer: It is both.
11. When performing maintenance on a process system with a high concentration of H_2S, what additional personal protective equipment should be worn by all craftsmen?
    Answer: Hose line supplied breathing air or self-contained breathing apparatus (SCBA).
12. What is one of the most important personal safety protective devices that workers should wear anytime they are working around or near equipment containing high concentrations of hydrogen sulfide?
    Answer: A personal hydrogen sulfide detection device equipped with audible, visual, and vibration alarms.
13. When the H_2S detector alarm sounds, how much time do we have before we should leave the area?
    Please mark the most correct answer:
    a. Everyone should leave the area immediately.
    b. 5 minutes
    c. 10 minutes
    d. It depends on the wind direction and windspeed.
    Answer: a. Everyone should leave the area immediately.
14. When unloading a truck of chemicals, the following should be in place to prevent from unloading chemicals into the wrong tank.
    Please mark all that apply:
    a. The facility unloading lines should be correctly labeled, color-coded, and identified for the specific chemical to be unloaded.
    b. Always verify the shipping documents that are provided by the truck driver to ensure the correct chemical is being delivered.
    c. The truck driver should know what chemical is being delivered.
    d. The unloading connections should be designed only for the chemical to be unloaded and incompatible with all other chemicals.

e. The truck driver and the facility representative, typically the operator, should both be present during the unloading.
f. The company should have a documented procedure for unloading the chemical and this procedure should be followed.
g. All of the above.

Answer: g. All of the above.

15. When gas testing a vessel for entry, why is it important to test for oxygen before testing for flammables or hydrogen sulfide?
Answer: The gas test machine may require adequate oxygen in order to provide an accurate LEL result.

16. When considering an excavation for a confined space, what depth triggers all the permit-required confined space entry requirements?
Answer: 4 feet

17. There are many requirements detailed by OSHA 29CFR 1910.146, the Confined Space Entry Standard. What are the three of the most important elements of the OSHA Confined Space Entry Standard?
Answer(s):
    1. Effective gas tests for oxygen, flammables, hydrogen sulfide, and any other toxic specific to the process involved.
    2. A designated standby attendant positioned at the point of entry who is in direct contact with the entrant and who can summon for help if rescue is required.
    3. A rescue plan with personnel and equipment readily available.

18. Please select the most likely places where serious levels of hydrogen sulfide may be expected and where precautions, including the use of self-contained breathing apparatus (SCBA), should be required.
Please mark all that apply:
    a. A grain silo with rotting grain in the bottom with sludge and water.
    b. A manure pit where manure from a cattle farm has been dumped.
    c. An old and abandoned oil and gas well with leaky pipes near the surface.
    d. A tank truck hauling water bottoms for disposal from crude oil storage tanks.
    e. A settling pond at a municipal sewerage treatment plant.
    f. A volcanic geyser in a national park.
    g. An open marshy field with decaying grass and other vegetation.
    h. A sewer manhole where underground sewerage transfer pumps are located.
    i. An empty railcar that recently carried a load of crude oil.
    j. A pit containing a pump that removes water draining from an adjacent landfill.
    k. None of the above.
    l. All of the above.

Answer: l. All of the above.

19. Your task is to add polymer to an asphalt tank through a gauging hatch on the top of the storage tank. What is the safest way to approach the gauging hatch?
    a. Approach the gauging hatch from the downwind direction.
    b. Approach the gauging hatch from the crosswind side of the hatch.
    c. Don a self-contained breathing apparatus (SCBA) and approach the hatch from the crosswind side of the hatch.
    d. Don a self-contained breathing apparatus (SCBA) and approach the hatch from the upwind side of the hatch.

    Answer: d. Don a self-contained breathing apparatus (SCBA) and approach the hatch from the upwind side of the hatch.
20. You are working with a crew that is using a hydro jet system to unclog the Quick Trip sewer line. After the crew successfully dislodged the obstruction, you are requested to enter the manhole to visually inspect the condition of the line. The manhole is 8 feet deep. How do you prepare to enter the manhole? Please mark the most correct answer.
    a. Gas test for oxygen and flammables, and if the tests are good, enter the manhole and perform the inspection.
    b. Ensure a gas test for oxygen, flammables, and hydrogen sulfide. Prepare a confined space entry permit documenting the gas test results, including a rescue plan, and a standby attendant.
    c. Prepare a confined space entry permit including a gas test for oxygen, flammables, and hydrogen sulfide. Prepare a rescue plan, including wearing a safety harness with a retrieval rope attached. Ensure a standby attendant who can notify rescuers.
    d. Don a self-contained breathing apparatus (SCBA) and enter the manhole and perform the inspection.

    Answer: c. Prepare a confined space entry permit including a gas test for oxygen, flammables, and hydrogen sulfide. Prepare a rescue plan, including wearing a safety harness with a retrieval rope attached. Ensure a standby attendant who can notify rescuers.

Chapter 8: End-of-Chapter Quiz (With Answers)

1. The Hierarchy of Controls was introduced by the National Safety Council in ________ as a safety system to reduce workplace ________ and reduce on-the-job injuries.
   Answer(s): 1950, hazards
2. US OSHA states the following about the Hierarchy of Controls: "The hierarchy of controls is a method of ________ and ________ safeguards to protect workers from hazards."
   Answer(s): identifying, ranking

3. How are the Hierarchy of Controls arranged?
   Please choose the most correct response.
   a. They are not arranged in any particular way.
   b. They are arranged in order from the least effective to the most effective.
   c. They are arranged in order from the most effective to the least effective.
   d. It depends entirely on the hazard being evaluated as to how they are arranged.
   Answer: c. They are arranged from the most effective to the least effective.
4. Please name the five controls in their appropriate order.
   Answer(s):
   1. Hazard elimination
   2. Substitution
   3. Engineering controls
   4. Administrative controls
   5. Personal protective equipment (PPE)
5. Which of the Hierarchy of Controls is the most effective?
   Answer: Elimination
6. Which of the Hierarchy of Controls is the least effective?
   Answer: Personal protective equipment (PPE)
7. What are the factors that make PPE the least effective control?
   Answer(s):
   - PPE may not be readily available at the time it is needed.
   - PPE may not be worn by the workers when it is needed.
   - PPE may not be worn correctly by the workers.
   - The workers may not respond to the alarms in a timely manner (for example, in the case of an H_2S detector).
   - The wearing of PPE may not be enforced by management.
8. Likewise, what makes Elimination the most effective of all the controls?
   Answer: If the hazard is eliminated, it can't present a hazard, or "What Your Don't Have, Can't Leak."
9. Which two of the controls are very difficult to implement in an existing process?
   Answer(s): Elimination and substitution
10. Which three of the controls can we easily make safety changes or improvements to address identified safety deficiencies or hazards?
    Answer(s): Engineering controls, administrative controls, personal protective equipment (PPE)

Chapter 9: End-of-Chapter Quiz (With Answers)

1. What is the purpose of the jobsite walkdown prior to opening process equipment?
   Answer: Every flange to be opened must be identified and properly tagged with a distinctive tag, identifying it as a point to be opened. This is to help ensure that only the correct flanges are opened.

2. When would it be acceptable to attempt to rescue or provide aid to a stricken coworker, one who collapsed while working on the jobsite?
   Answer: Only if I am properly trained in rescue techniques and equipped with the proper personal protective equipment, such as a self-contained breathing apparatus (SCBA), and a backup person is also equipped with SCBA.
3. What are the three key elements of a successful confined space entry program?
   Answer(s):
   1. Successful gas test of the confined space (oxygen, flammables, and hydrogen sulfide, and other toxics specific to the confined space being entered).
   2. Assigned and dedicated standby person at the point of entry who can notify for rescue if needed.
   3. Detailed and ready rescue plan with equipment on standby.
4. What is the proper response if a coworker collapses while on the jobsite?
   Answer: Instead of rushing in to provide assistance, we should immediately call for emergency response and then standby to direct the rescue personnel to the victim.
5. What does a successful gas test for oxygen look like?
   Answer: 19.5% oxygen or more, but less than 23% oxygen.
6. What is the ideal approach for evacuating vapors from the worksite when opening process equipment?
   Answer: Forced air ventilation should be used wherever possible to help evacuate the toxic vapors away from the workers.
7. What type of personal protective equipment should the workers use when opening process equipment?
   Answer: It is important that workers are provided with and wear appropriate personal protective equipment (PPE) like chemical safety goggles, full-body protective suits, and a self-contained breathing apparatus (SCBA) when warranted by the gas test. There should be on-going air quality monitoring by each worker with a hydrogen sulfide detector.
8. The purpose of the standby attendant during a confined space entry is to remain in direct __________ with the entrant(s) throughout the confined space entry. The attendant must be prepared and ready to summon for __________ in an event this is required by the entrant(s).
   Answer(s): contact, rescue
9. A detailed workplan incorporating jobsite safety when working around or on systems and equipment containing, or that has contained, high concentrations of hydrogen sulfide can save __________ and prevent tragedy that affects families for a __________, and beyond.
   Answer(s): lives, lifetime
10. A good place to start addressing our long history of hydrogen sulfide poisoning incidents is with __________ and __________.
    Answer(s): training, education

11. Initial hydrogen sulfide awareness __________ should be provided for all personnel who may be required to work in areas where H_2S is processed or handled, including an annual ______________ training.
Answer(s): training, refresher
12. Everyone who works at a facility where hydrogens sulfide is produced, stored, or handled in high concentrations should be equipped with a personal H_2S ___________ that has both __________ and __________ alarm capability.
Answer(s): detector, audible, visual
13. Prior to entry into an excavation more than _______ feet in depth, the atmosphere should be tested ________ to entry.
Answer(s): 4, prior
14. To ensure success, the requirement to wear a personal H_2S detector should be formally adopted by the facility and enforced as a safety rule that wearing the detector is a condition of continued ______________ at the facility.
Answer: employment
15. Facilities that have processing equipment with high concentrations of H_2S should be equipped with fixed H_2S sensors with __________ and ____________ alarms in the event of an H_2S release.
Answer(s): audible, visual
16. All personnel working at the facility must be trained in the alarm sound and __________ to take in the event of a detected release, including immediate evacuation of the area by moving crosswind (never ________ or __________).
Answer(s): actions, upwind, downwind
17. The flash point for hydrogen sulfide is __________, and the flammable range is __________% to ___________%, making it extremely ______________.
Answer(s): −116 °F (−82 °C), 4.3, 45.5, flammable

Appendix 2

Documented Cases of Hydrogen Sulfide Poisoning

30 Years of Data 1993–2023/30 Years of Tragedy

Extracted from the Occupational Safety and Health Administration (OSHA) Database

October 10, 2024: US Chemical Safety and Hazard Investigation Board (CSB) Update Report dated March 2025 (OSHA Report still pending as of this writing) (2 Fatalities, 13 Evaluations at the hospital (local media reported 35 injured), broad areas of Deer Park and Pasadena, TX requested to shelter in place).

On October 10, 2024, at 4:23 p.m., approximately 27,000 pounds of hydrogen sulfide gas were released during a maintenance activity at the PEMEX Deer Park Refinery in Deer Park, Texas. The release fatally injured two contract workers. Officials in the neighboring cities of Deer Park and Pasadena issued shelter-in-place orders that lasted several hours. The contract pipefitters opened a flange on Unit AR-7 instead of Unit AR-6 which had been depressured and prepared for maintenance (the wrong unit). The flange that was opened contained hydrogen sulfide at 90% (900,000 ppm) while operating at normal flow and pressure.

May 29, 2024, OSHA Database (1 Fatality, 3 Injuries) (Attempted Rescue)

At 3:22 p.m. on May 29, 2024, Employee #1 was attempting to rescue a fellow employee who had been overcome by hydrogen sulfide (H_2S) gas in an underground pump house. Prior to this incident, multiple coworkers had been working

Hydrogen Sulfide (H_2S) Hazards: Risk Recognition and Prevention, First Edition. M. Darryl Yoes.

in the pump house and were exposed to H_2S gas emitted from a pipe containing water with by-products from sugar beets, dirt, mud, and chemicals used in water treatment. The victim became overwhelmed by the H_2S gas, and Employee #1 was also overexposed while attempting the rescue. The employee lost consciousness and was later found unresponsive.

May 6, 2024, Italian Media Report (5 Fatalities, 1 Worker in intensive care)

Five workers died on Monday afternoon while working in a sewer connected to a water treatment plant in the town of Casteldaccia near Palermo, Sicily, according to firefighters who recovered the bodies. A sixth worker was seriously injured and in intensive care, local officials said. Girolamo Bentivoglio, chief of firefighters in Palermo, said the workers had breathed in hydrogen sulfide, a gas often present at waste treatment plants that is toxic in high concentrations. Levels were so high at the site where the workers were killed, "that fatality is immediate. Had they adopted all the precautions, all the safety measures, including personal protective equipment, we wouldn't have been faced with this scenario" Mr. Bentivoglio said. Prime Minister Giorgia Meloni of Italy called for "full light to be shed on this tragedy" in a post on social media site X.

December 2023 OSHA Database: 1 Fatality

Mixing chemicals resulted in H_2S

At 11:00 a.m. on December 19, 2023, an employee who performed maintenance work for petroleum products wholesaler was called to check on a plumbing issue at a gas station. The employee left the site and returned at 3:30 p.m. to complete the job. The employee went into the bathroom to continue working on the plumbing issue. He removed the toilet and poured muriatic acid into the drains. The employee was found unconscious at 4:30 p.m. When emergency responders arrived on site, their personal hydrogen sulfide (H_2S) monitors began to alarm even prior to entry into the bathroom where the employee was located. The employee was transported to a medical center and then transferred to another medical center. On December 30, 2023, the employee was taken off of life support and died. The cause of death was determined to be H_2S and carbon monoxide (CO) poisoning.

https://www.osha.gov/ords/imis/accidentsearch.accident_detail?id=163060.015

July 2023. OSHA Database: 1 Fatality

At 12:45 p.m. on July 3, 2023, an employee climbed down a fixed ladder within a manhole structure to retrieve a stuck tool caught on a ladder rung. As the employee began to climb out of the manhole opening, the employee lost consciousness and fell approximately 30 feet to the bottom of the manhole. The employee died as a result of toxic, fatal levels of hydrogen sulfide gas exposure and blunt force injuries to the head, resulting from the fall.

https://www.osha.gov/ords/imis/accidentsearch.accident_detail?id=157755.015

February 24, 2023. OSHA Database: 1 Fatality

Attempted Rescue

At approximately 12:00 p.m. on February 24, 2023, an employee was performing well site pumping operations for an oil and gas producer. A coworker had been trying to contact the employee unsuccessfully, so he went to the oil and gas well site where the employee was working to check on him. The coworker found the employee unresponsive in the small 9-foot by 10-foot well shack. The door had been closed when the coworker arrived. The well site is a known to have hydrogen sulfide present in the air at times. The coworker observed that a hose was attached to a valve on the discharge side of a pump. The pump is used to transfer fluid from a tank that is used to separate natural well gases from fluid (condensate and produced water) prior to the gases going to a flare. The coworker observed an overturned bucket in the shack with a small puddle of fluid at the end of the hose. The employee was using the hose and bucket in an attempt to clear a blockage from the pump. The coworker pulled the employee from the space and summoned emergency response. He conducted CPR while he waited for medical help to arrive. The coworker reported that the employee had foam in his mouth and blood was coming out of the employee's mouth when he performed CPR. The employee also had chemical type burns. The employee was transported to the hospital, where he was pronounced dead. The employee was overcome by high levels of flammable gases, including potential hydrogen sulfide overexposure.

https://www.osha.gov/ords/imis/accidentsearch.accident_detail?id=154242.015

February 3, 2023 Local Media Report 1 Fatality

As yet to be confirmed, a worker from Texas was reportedly killed in died in February 2023 from exposure to hydrogen sulfide at Suncor Energy's Commerce City oil refinery, his family alleges in a new wrongful death lawsuit.

September 2022. OSHA Database: 1 Fatality

At 9:00 p.m. on September 19, 2022, an employee was pumping wastewater out of a drain sludge containment area when, after the operation was complete, he lifted the suction hose and fell to the ground. The employee died due to acute thiosulfate intoxication from inhalation of hydrogen sulfide.

https://www.osha.gov/ords/imis/accidentsearch.accident_detail?id=149915.015

August 2022. OSHA Database: 1 Fatality, 3 Injuries (Released same day)

Attempted Rescue

At 7:00 a.m. on August 22, 2022, Employee #1 was operating an excavator at a hazardous waste treatment and disposal facility. Employee #1 was working in the Solids Processing Unit and was following a batch sheet recipe to mix neutralizing agents with industrial waste. Employee #1 went to the Liquids Processing Unit and asked a coworker to send spent sulfuric acid "pickle liquor" to the bay where he was working. Employee #1 returned to his excavator so he could mix the batch when the "pickle liquor" was sent. Moments after the "pickle liquor" was sent, Employee #1's excavator swung into the sidewall of the Solids Processing Unit. Employee #2 rushed over to see what happened and found Employee #1 unconscious. When Employee #2 attempted to climb up into the excavator to help Employee #1, he became lightheaded and lost consciousness. However, Employee #2 was able to regain consciousness and go for help and 9-1-1 was contacted. Additional coworkers arrived to assist Employee #1, including Employees #3 and #4. Employee #1 was transported to the hospital and treated for respiratory chemical exposure, but was declared brain dead on August 24, 2022, and died the following day. Employees #2, #3 and #4 were observed at the hospital and released the same day with no diagnosed injuries. It is unknown what chemical the employees were exposed to but was suspected to be hydrogen sulfide.

https://www.osha.gov/ords/imis/accidentsearch.accident_detail?id=149136.015

June 2022 OSHA Database: 2 Fatalities

At 8:00 a.m. on June 26, 2022, two employees were emptying and cleaning an open topped frack tank (Tank #4) at an oil and gas services site. There were no eyewitnesses and the employees were working at remote site by themselves, so few details are known about what occurred. At the time of the investigation, hydrogen sulfide (H_2S) gas was detected at levels between 3 parts per million (ppm) and 5 ppm, which is non-lethal. The two employees died. But, it is not possible to estimate what the H_2S levels might have been when the employees were killed. The tank contained approximately 2 feet of water/oil. Autopsy results are pending.

https://www.osha.gov/ords/imis/accidentsearch.accident_detail?id=147970.015

June 2022 OSHA Database: 1 Fatality

Attempted Rescue

On June 7, 2022, an employee working as an environmental engineer was replacing two failed pumps in a pump station. Two coworkers connected the new pumps to electric and the employee lowered the pumps into the vault. It was noted that the pumps weren't working correctly and water was pumped from the space. The employee was allegedly overcome by hydrogen sulfide (H_2S), fell into the vault and was killed. A coworker working for a contractor called for help after observing the employee fall and was also found in the vault.

https://www.osha.gov/ords/imis/accidentsearch.accident_detail?id=146891.015

May 2022 OSHA Database: 1 Fatality, 1 Injury (hospitalization)

H_2S Explosion/Fire

At 9:45 a.m. on May 14, 2022, Employee #1, Employee #2, and Employee #3 were working near a trailer tank when a torch to seal leaks ignited hydrogen sulfide (H_2S), creating a fireball. Employee #1, Employee #2, and Employee #3 were hospitalized to treat second- and third-degree burns. Employee #1 died from his injuries on May 23, 2022.

https://www.osha.gov/ords/imis/accidentsearch.accident_detail?id=146481.015

October 2021 OSHA Database: 0 Fatalities, 1 Injury (hospitalization)

H_2S Explosion/Fire

At 6:45 p.m. on October 28, 2021, an employee was circulating a tank of hydrogen sulfide when a burner box, which was 15 feet away, was turned on and the hydrogen sulfide that was in the atmosphere ignited. The employee was hospitalized to treat second and third-degree burns to his hands and face.

https://www.osha.gov/ords/imis/accidentsearch.accident_detail?id=145747.015

September 2021 OSHA Database: 2 Fatalities

On September 12, 2021, two employees entered a pit to work on unjamming an auger. The employees removed the cover of the auger, at which time a large amount of hydrogen sulfide was released. The employees were both overcome by the gas. The employees were found over an hour later face down in the pit and were pronounced dead at the scene.

https://www.osha.gov/ords/imis/accidentsearch.accident_detail?id=139143.015

August 2021 *Miami Herald News Article 3 Fatalities (3 Brothers)

Three brothers die after passing out in manure pit at Ohio farm, authorities say BY MITCHELL WILLETTS August 11, 2021 12:59 PM. At the time of the Tuesday afternoon accident, the brothers — Gary, Todd and Brad Wuebker — were performing maintenance on a pump inside the storage pit, the Mercer County Outlook reported. Manure pits are known to produce a wide range of toxic gases. St. Henry firefighters received a call for help around 12:30 p.m., the paper reported, and when they arrived at the livestock operation, all three men were unconscious in the pit. First responders used ropes and ladders to pull the men out, WLIO reported.

Read more at: https://www.miamiherald.com/news/nation-world/national/article253418144.html#storylink=cpy

August 2021 OSHA Database: 1 Fatality

At 12:00 noon on August 27, 2021, an employee entered a tanker trailer to power wash the interior of the tanker. The tanker had contained animal parts, and was sitting outside in the high summer temperatures. The employee was found unconscious inside the tanker and emergency services were notified. Upon arrival, the fire department detected hydrogen sulfide at twice the level considered to be immediately dangerous to life and health at the open manway on the top of the tanker. No air monitoring had been done prior to entry into the tanker.

No attendant was in place, and there was no retrieval system set up to enable non-entry rescue of the employee from outside the tanker. The employee died due to asphyxiation.

https://www.osha.gov/ords/imis/accidentsearch.accident_detail?id=138824.015

June 2021 OSHA Database: 1 Fatality, 1 Injury (Hospitalization)

Attempted Rescue

At 11:00 a.m. on June 8, 2021, Employee #1 washed the inside of a tanker to clear chemical residue from the inside of the tanker. At some point, Employee #1 lost consciousness and was found by Employee #2, who attempted to rescue Employee #1 from the tanker. A coworker heard Employee #2 in distress and called for emergency medical assistance. Employee #1 was killed and Employee #2 was hospitalized due to exposure to hydrogen sulfide.

https://www.osha.gov/ords/imis/accidentsearch.accident_detail?id=136167.015

June 2021 OSHA Database: 1 Fatality

At 12:30 p.m. on June 7, 2021, an employee entered a sewer manhole to estimate the amount of materials needed to perform a manhole repair when the employee was overcome by high levels of hydrogen sulfide at the base of the manhole shortly after entry. The employee did not perform atmospheric monitoring and had not donned his harness with rescue tripod before entry into the confined space. A four gas monitor and rescue equipment were located in the employee's work vehicle. Coworker #1 (attendant) and Coworker #2 (helper) were not trained or equipped for performing a confined space entry rescue. A volunteer fire department responded within 8 minutes and performed the confined space entry rescue shortly after arrival. The employee was killed by overexposure to hydrogen sulfide.

https://www.osha.gov/ords/imis/accidentsearch.accident_detail?id=136124.015

May 2021 OSHA Database: 2 Fatalities

At 1:00 p.m. on May 25, 2021, Employee #1 and Employee #2 were conducting maintenance work on a pump above a leachate vault. A coworker dropped a cellphone inside the vault and left to look for a tool to retrieve the cellphone.

Employee #1 and Employee #2 entered the vault to retrieve the phone and were overcome by methane and hydrogen sulfide gases and were killed.
https://www.osha.gov/ords/imis/accidentsearch.accident_detail?id=136332.015

July 2020 OSHA Database: 2 Fatalities, 2 Treated and Released

Attempted Rescue

At 4:41 p.m. on July 24, 2020, Employees #1 and #2 set a 32 foot ladder into an eight and a half foot manhole that was connected to the main city sewer line station. Employee #1 entered the manhole and was immediately overcome by hydrogen sulfide gas, Employee #2 entered and was also overcome by the gas. Employees #3 and #4 entered to rescue the employees and were able to rescue them from the manhole. Employee #1 was killed by the hydrogen sulfide exposure, Employee #2 later died at the hospital and Employees #3 and #4 were treated and released.
https://www.osha.gov/ords/imis/accidentsearch.accident_detail?id=128141.015

July 2020 OSHA Database: 2 Fatalities

Attempted Rescue

At 11:30 a.m. on July 9, 2020, Employees #1 and #2 and their foreman, employed by a waste management company, were installing a closed circuit television (CCTV) into an 8-inch sewer line to enable inspections in the manhole. They did not test the atmosphere before attempting to enter the manhole, which met the definition of confined space, and ventilation equipment was not available. Employee #1 entered the 12-foot deep manhole to remove a plug that controlled the flow of water and was overcome by the release of water and sewer gases that included toxic hydrogen sulfide. Employee #2 entered the manhole to provide assistance, was overcome by the sewer gases, and fell from the rope ladder he was using to the bottom of the manhole. The foreman was unable to rescue the employees. A 3M Confined Space Tripod system was onsite but not set in place. Employee #1 and Employee #2 both died in the manhole.
https://www.osha.gov/ords/imis/accidentsearch.accident_detail?id=127680.015

June 2020 OSHA Database: 0 Fatalities, 1 Injury (Hospitalized)

At 10:00 a.m. on June 18, 2020, an employee was servicing a gas well, checking the fluid level. The employee was exposed to hydrogen sulfide gas and was hospitalized to treat unspecified injuries. The employee was not trained and was not wearing any PPE.
https://www.osha.gov/ords/imis/accidentsearch.accident_detail?id=128174.015

May 2020 OSHA Database: 1 Fatality, 1 Injury (Not Hospitalized)

Attempted Rescue

At 9:00 a.m. on May 29, 2020, Employees # 1 and #2 coworkers discussed entry into a manhole to clear a blockage. Employee # 1 entered into the manhole without personal protective equipment, without training, and without performing any air quality testing. Employee # 1 successfully placed the line to clear the blockage and began to climb out. Before he could exit, he was overtaken, likely by hydrogen sulfide, an oxygen deficient atmosphere and lost consciousness. The manhole contained approximately two feet of standing water. When the others realized that Employee # 1 was unconscious, Coworker #1 tied a ratchet strap around a fence adjacent to the manhole entrance and around his waist. As Coworker #2 contacted emergency services, Coworker #1 entered the 17 foot deep manhole. Coworker #1 reached Employee # 1, lifted him out of the water, but was unable to pull him out of the manhole. Coworker #1 began climbing out of the manhole but lost consciousness prior to reaching the top. Coworker #2 pulled Coworker #1 up several feet allowing Coworker #1 access to fresh air. He regained consciousness and began climbing out of the manhole with assistance from Coworker #2. Employee # 1 could not be rescued. Coworker #1 was taken to the hospital for assessment of his injuries (cuts and bruises). After more than one hour and positive pressure ventilation had been implemented in the manhole, atmospheric monitoring performed by the fire department indicated that oxygen levels were at 20.4% and hydrogen sulfide was present at 5.5– 6 parts per million. Employee # 1 was killed. Coworker #2 was provided basic screening for injury and illness by emergency services at the site as he had momentarily entered the manhole during his rescue of Coworker #1. He was not hospitalized.

https://www.osha.gov/ords/imis/accidentsearch.accident_detail?id=126541.015

February 2020 OSHA Database: 0 Fatalities, 1 Injury (Hospitalization)

At 5:30 p.m. on February 28, 2020, Employee #1 was changing the drain hose from one bucket to another. The fluid being drained released hydrogen sulfide gas, causing the employee to lose consciousness. The employee was hospitalized and treated for exposure to hydrogen sulfide gas.

https://www.osha.gov/ords/imis/accidentsearch.accident_detail?id=124487.015

January 2020 OSHA Database: 0 Fatalities, 1 Injury (Hospitalization)

At 8:45 a.m. on January 5, 2020, an employee was exposed to hydrogen sulfide while gauging a tank. The employee experienced upper respiratory irritation, dizziness, and disorientation. The employee was hospitalized.

https://www.osha.gov/ords/imis/accidentsearch.accident_detail?id=123068.015

December 2019 OSHA Database: 0 Fatalities, 2 Injuries (1 Hospitalized)

Attempted Rescue

At 11:00 p.m. on December 16, 2019, Employee #1 and Employee #2, both chemical technicians working for a fertilizer manufacturer, were "popping-the-top" (opening the top hatch) of sulfur train car. Employee #1 was performing the work while Employee #2 served as the "spotter," relaying the instructions of employee #1 to a coworker (locomotive operator). Employee #2 lost consciousness causing Employee #1 to remove his respirator to help Employee #2. As a result Employee #1 suffered pulmonary (chemical) edema and was hospitalized. Employee #2 sustained hemorrhaging in his eyes due to breathing stress and required medical treatment without hospitalization.

https://www.osha.gov/ords/imis/accidentsearch.accident_detail?id=122638.015

December 2019 OSHA Database: 0 Fatalities, 1 Injury (Hospitalization)

At 6:35 p.m. on December 11, 2019, an employee was removing bolts around a railcar hatch. A rapid release of hydrogen sulfide gas from the railcar somehow entered the employee's respirator mask. The employee suffered respiratory irritation or arrest brought on by acute exposure to hydrogen sulfide gas. The employee was hospitalized overnight and then released with no work restrictions.

https://www.osha.gov/ords/imis/accidentsearch.accident_detail?id=122599.015

October 2019 OSHA Database: 2 Fatalities (1 Fatality Reported by OSHA)

AT 7:00 p.m. on October 26, 2019, an employee entered the pump house to investigate an alarm. The employee was overcome by hydrogen sulfide gas when Pump #1 produced water containing hydrogen sulfide and contaminated the atmosphere.

The employee died from asphyxiation. **Note: * The employee's wife was also killed when she attempted to help her husband. Not reported by OSHA.**

https://www.osha.gov/ords/imis/accidentsearch.accident_detail?id=121136.015

October 2019 OSHA Database: 0 Fatalities, 1 Injury (Hospitalization)

At 1:30 pm on October 17, 2019, an employee working for a meat processing plant was standing on the top of the tanker truck. The employee was using a tape measure gauging the depth of the waste blood product that was being pumped into the tanker for an adjacent rail tanker. A coworker found the employee lying on his stomach with his head and right arm inside the manhole of the tanker truck. The employee had lacerated his forehead but no other visible injuries were observed. The employee was unresponsive, was not breathing and had no pulse at the time. The employee was hospitalized which were attributed to unknown causes at the time the narrative was submitted. However, a measurement was taken with a four gas meter at the hatch opening shortly after the employee was discovered, which showed a hydrogen sulfide (H_2S) level well above the ceiling level of 50 ppm.

https://www.osha.gov/ords/imis/accidentsearch.accident_detail?id=121109.015

September 2019 OSHA Database: 0 Fatalities, 1 Injury (Hospitalization)

At 10:30 a.m. on September 30, 2019, an employee was performing maintenance on a storage tank battery for oil and gas operations support. The employee was on a crew which was working on a line which was not pressurizing properly. As the employee was working on the line, the pop off valve relieved causing the employee to be struck by released hydrogen sulfide. The employee fell off the tank, rupturing his kidney and fracturing his left fibula and right hip. The employee was hospitalized.

https://www.osha.gov/ords/imis/accidentsearch.accident_detail?id=120777.015

May 2019 OSHA Database: 1 Fatality

At 4:00 p.m. on May 24, 2019, Employee #1, a farmer, entered a 1,500 gallon bulk milk tank to clean it. He was using a paste of cleaning granules and well water. The employee died within 20 minutes of entry into the tank due to the inhalation of excess concentrations of hydrogen sulfide (H_2S) that evolved from well water that was contaminated with naturally-occurring hydrogen sulfide. The lack of ventilation in the tank trapped the gas. Lab tests of the vapor space above a sample of the well water showed more than 1,000 ppm H_2S versus the allowable 8-hour exposure limit of 10 ppm. The cleaning granules did not play a role in the event.

https://www.osha.gov/ords/imis/accidentsearch.accident_detail?id=116597.015

May 2019 OSHA Database: 0 Fatalities, 1 Injury (Hospitalization)

Attempted Rescue

At 9:15 a.m. on May 8, 2019, Employee #1 and a coworker, town employees, were engaged in repair/maintenance of an underground pipeline. As his coworker lowered Employee #1 into a manhole, Employee #1 lost consciousness and was retrieved by the coworker. Employee #1 was transported to the hospital and admitted for treatment. After the incident, air testing revealed high levels of hydrogen sulfide gas and carbon monoxide.

https://www.osha.gov/ords/imis/accidentsearch.accident_detail?id=116131.015

April 2019 OSHA Database: 1 Fatality

At 3:00 a.m. on April 10, 2019, an employee was entering a vault manhole in order to gain access to the grease trap manhole. The employee entered the manhole vault and was overcome by the Hydrogen Sulfide fumes, fell into the vault, and landed on top of the grease trap. According to the medical examiner, the employee was killed the fumes within the vault.

https://www.osha.gov/ords/imis/accidentsearch.accident_detail?id=115356.015

January 2019 OSHA Database: 0 Fatalities, 1 Injury (Hospitalization)

At 6:15 p.m. on January 24, 2019, an employee was transferring water from an oil gas well site when he experienced some dizziness brought on by exposure to toxic gases such as hydrogen sulfide. The employee was hospitalized.

https://www.osha.gov/ords/imis/accidentsearch.accident_detail?id=113293.015

November 2018 OSHA Database: 0 Fatalities, 1 Injury (Hospitalization)

At 6:00 p.m. on November 20, 2018, an employee was lining up a water injection pump and the misalignment of valves caused a release of NAPSA and Hydrogen Sulfide. The employee was overcome when he inhaled the fumes and was hospitalized.
https://www.osha.gov/ords/imis/accidentsearch.accident_detail?id=111211.015

September 2018 OSHA Database: 1 Fatality

At 5:00 a.m. on September 27, 2018, the employee was unloading and washing a truck tanker with water that had contained poultry blood. The employee was found unresponsive on top of the tanker with his head over the tanker opening. The employee was killed from exposure to ammonia and hydrogen sulfide vapors from the tanker truck.
https://www.osha.gov/ords/imis/accidentsearch.accident_detail?id=109551.015

August 2018 OSHA Database: 0 Fatalities, 1 Injury (Hospitalization)

At 3:00 p.m. on August 20, 2018, an employee was bulldozing sludge material. While bulldozing, the employee was overcome from inhalation of hydrogen sulfide gas and became unconscious with respiratory distress. The employee was hospitalized.
https://www.osha.gov/ords/imis/accidentsearch.accident_detail?id=108561.015

July 2018 OSHA Database: 1 Fatality

At 2:25 p.m. on July 5, 2018, Employee #1 was installing a sewer line. The employee was overcome by the hydrogen sulfide and methane gas in the trench and died.
https://www.osha.gov/ords/imis/accidentsearch.accident_detail?id=107089.015

May 2018 OSHA Database: 1 Fatality

At 8:15 p.m. on May 25, 2018, Employee #1 was power washing the outside of the semi-truck tanker trailer. The employee entered the tanker trailer to wash the inside and was not found for two hours. The fire department was called to rescue the employee. Atmospheric monitoring found atmospheric levels of hydrogen sulfide at 100 ppm and hydrogen cyanide at 30 ppm. No written evidence of atmospheric monitoring was available following the employee's recovery from the space, and no ventilation of the space prior to or during the entry was performed. No attendant or entry supervisor was assigned to the entry. Employee #1's death was determined to be chemical asphyxia by vitiated atmosphere with hydrogen sulfide and hydrogen cyanide gasses.

https://www.osha.gov/ords/imis/accidentsearch.accident_detail?id=105985.015

April 2018 OSHA Database: 1 Fatality

At 11:30 p.m. on April 14, 2018, an employee was performing maintenance on the acid cat and replaced the injection methanol pump. The employee was overcome by a high concentration of hydrogen sulfide and was asphyxiated.

https://www.osha.gov/ords/imis/accidentsearch.accident_detail?id=104780.015

February 2018 OSHA Database: 0 Fatalities, 1 Injury (Hospitalization)

At 11:15 a.m. on February 9, 2018, an employee was working on the rig near the shakers and overcome from hydrogen sulfide inhalation and overexposure. The employee was hospitalized.

https://www.osha.gov/ords/imis/accidentsearch.accident_detail?id=107039.015

January 2018 OSHA Database: 1 Fatality

There are three inspections associated with this investigation. They are numbers 1299111, 1294521, and 1293967. This report is based on the information in report number 1293967. At 4:30 p.m. on January 31, 2018, Employee #1 was working for a building finishing contractor. He was at a jobsite using a welding rig. He was repairing metal hog pens within a hog confinement. This rig was manufactured by Hobart. It consisted of a generator/welder with a 305 cc (18.6 cubic inch) 19.7 Newton-meter (14.50 pound-foot) gross torque Brigg and Stratton Series 1450 gasoline combustion engine. Earlier that day, at approximately 3:30 p.m., Coworker #1 was working at the same facility. He was working for another contracting company. He had completed his shift for the day and went to use the restroom located within the jobsite's breeding room. He did not notice anything unusual upon entering the restroom. Shortly after he sat down on the stool, his heart began racing and he began having a difficult time breathing. Coworker #1 quickly exited the restroom and went back out into the breeding room, where he collapsed onto the ground. He was taken by Coworker #2 to the hallway, where his symptoms subsided. Coworker #1 stated that he did not know what had happened and thought he might have had a heart condition. With the assistance of other coworkers, Coworker #1 exited the area. At approximately 4:00 p.m. Employee #1 went to use the same restroom. He was in the restroom for approximately 20 minutes before his coworkers noticed that he had not returned. His coworkers, at the direction of company, went looking for him. A short time later, they found him in the restroom. He was unresponsive. He was sitting on the stool and slumped over an adjacent sink. He was foaming at the mouth with labored breathing. He had been potentially overcome with hydrogen sulfide, in combination with carbon monoxide from his welding activities. He was hospitalized as a result of this exposure to those gases. On February 2, 2018, he was pronounced dead. He had been welding from 6:00 a.m. until the time he went to use the restroom at 4:00 p.m., with only a short lunch break. The exposure to carbon monoxide during this activity, along with the presence of high levels of hydrogen sulfide, was thought to be the probable cause of this fatality.

https://www.osha.gov/ords/imis/accidentsearch.accident_detail?id=102977.015

November 2017 OSHA Database: 0 Fatalities, 1 Injury (Hospitalization)

On November 20, 2017, an employee was blowing down a transmitter and was exposed to hydrogen sulfide. The employee fell down in the boom lift basket and onto a steam hose, suffering second and third degree burns to his inner thighs. The employee was hospitalized.

https://www.osha.gov/ords/imis/accidentsearch.accident_detail?id=101065.015

November 2017 OSHA Database: 1 Fatality

At 12:30 p.m. on November 14, 2017, a 3 person crew was returning the well to service following repairs to a wellbore casing nipple that was performed by a welding contractor. A welding contractor repaired a crack on the casing nipple of the oil well and then departed. The three person crew had set a "packer" on the oil well to prevent oil and hydrogen sulfide from escaping. Thirty barrels of salt water was inserted on top of the packer. After the crew pulled the "packer" out, oil and gas started discharging from the well. At that time, hydrogen sulfide monitors located in a portable steel structure used for tools, as a lunchroom, and as an office & referred to as the Doghouse were activated and an alarm was triggered. The "Doghouse" was located approximately 60 feet from the well. The crew realized there was a propane heater with a pilot light burning in the Doghouse. One of the employees ran to turn off the propane tank outside the structure while another ran inside the Doghouse. The wind blew the hydrogen sulfide into the Doghouse which found an ignition source, presumably the pilot light, and a fire erupted. The employee was wearing flame retardant clothing; However, due to the intensity of the fire, it did not prevent injury. The employee died November 16, 2017, as a result of burn injuries.

https://www.osha.gov/ords/imis/accidentsearch.accident_detail?id=100750.015

August 2017 OSHA Database: 0 Fatalities, Hospitalized (Released same day)

At 10:45 a.m. on August 2, 2017, Employee #1 was working with a crew that was using a hydrojet system to unclog the Quick Trip sewer line. After the crew successfully dislodged the obstruction, the employee entered the active sewer manhole to visually inspect the condition of the line. The employee collapsed and fell to the bottom of the 7.5 foot deep manhole. The fire department responded and performed atmospheric testing and found elevated hydrogen sulfide levels at approximately 193 ppm. Employee #1 was transported by ambulance to the hospital, where he was treated and released the same day.

https://www.osha.gov/ords/imis/accidentsearch.accident_detail?id=97840.015

July 2017 OSHA Database: 3 Fatalities

Attempted Rescue

At 11:00 a.m. on July 7, 2017, Employee #1was attempting to dislodge a 24 inch rubber plug from a 2 foot diameter sewer pipe located inside a 24 foot deep wet well. The workers were outside the well pulling on a 1/4-inch nylon rope that was attached to the 24-inch diameter plug. The plug was lodged inside a T-shaped PVC fitting from the force of the waste water emptying into the well. Without conducting any atmospheric testing of the work space, Employee #1 climbed down the ladder with a crowbar to dislodge the deflated 24 inch diameter rubber plug, which was about 8 feet below the top of the well. He had difficulty releasing the plug with the crowbar and started to make his way up the ladder. He lost consciousness when he was about 2 feet from the top of the well and fell into the 24 foot deep well. Employee #2 descended down the ladder to provide emergency rescue, but lost consciousness and went underwater. The waste water level was about 3 feet deep at this point. Employee #3 climbed down the ladder to provide emergency rescue, but lost consciousness as well. All three workers were asphyxiated by hydrogen sulfide (H_2S) gas.

https://www.osha.gov/ords/imis/accidentsearch.accident_detail?id=97526.015

June 2017 OSHA Database: 2 Fatalities

Father and Son Killed

At 2:00 p.m. on June 13, 2017, two employees, a father and son team, were replacing a pump. The son was overcome by hydrogen sulfide gas and became unconscious. As the father retrieved his son from the pump lift station, he fell and a suffered blunt force trauma injury. Both father and son died.

https://www.osha.gov/ords/imis/accidentsearch.accident_detail?id=96339.015

February 2017 OSHA Database 1 Fatality

At 9:30 a.m. on February 3, 2017, an employee was cutting out old sewer pipes under a residential home when he was overcome by hydrogen sulfide sewer gas.

https://www.osha.gov/ords/imis/accidentsearch.accident_detail?id=92831.015

January 2017 OSHA Database: 1 Fatality

At 8:30 a.m. on January 16, 2017, an employee was descending a ladder into a manhole to repair a leaking pipe. The employee was overcome by gasses inside the manhole and was asphyxiated by hydrogen sulfide and methane gas.

https://www.osha.gov/ords/imis/accidentsearch.accident_detail?id=96757.015

January 2017 Separate OSHA Report: 3 Fatalities, 3 Injures

(This report was in an OSHA Report but was not found in the OSHA Database)

Attempted Rescue

A 34-year-old pipe layer entered the manhole – a confined space – and quickly became unresponsive. A 49-year-old laborer entered the hole and attempted to rescue the first employee. After the second employee also became unresponsive, a 24-year-old equipment operator followed to help his fallen coworkers. All three men died. Post-incident atmospheric testing in the manhole revealed lethal levels of hydrogen sulfide and carbon monoxide. Two other employees and a volunteer firefighter were also exposed to the toxic gases in the manhole during rescue attempts but survived.

https://www.osha.gov/news/newsreleases/region4/07182017

January 2017 OSHA Database: 0 Fatalities, 1 Injury (Hospitalization)

At 9:45 a.m. on January 9, 2017, an employee was working for a firm engaged in crude petroleum and natural gas extraction. He was testing the level in a well through the use of an "EKO gun." The employee connected the EKO gun to shoot the level ("check the well level"). When he cranked a two-section ball valve, a few threads slipped out. The valve split open and sprayed gas onto the employee. The well gases included hydrogen sulfide, methane, and so forth. The employee was overcome by the gases, but he was not suffocated. He was hospitalized.

https://www.osha.gov/ords/imis/accidentsearch.accident_detail?id=92163.015

October 2016 OSHA Database: 2 Fatalities

Attempted Rescue

At 11:00 a.m. on October 18, 2016, two employees were removing a failed sump pump from a wet well and both were killed by hydrogen sulfide exposure.

https://www.osha.gov/ords/imis/accidentsearch.accident_detail?id=89855.015

April 2015 OSHA Database: 1 Fatality

At approximately 4:45 p.m. on April 22, 2015, Employee #1 was working at a polymer plant and adding bags of polymer cross linker chemicals to an asphalt tank. He was on top of a platform for asphalt Mix-1 tank. As he added the chemical, he was overexposed to the inhalation of hydrogen sulfide fumes emanating from the tank and was killed.

https://www.osha.gov/ords/imis/accidentsearch.accident_detail?id=76282.015

January 2015 OSHA Database: 1 Fatality

At approximately 10:30 p.m. on January 19, 2015, an employee was responding to an alert from a warning panel about high level of fluid in the tank. The employee worked on the system earlier in the day and was responding to see what was wrong with the system. The employee was working on electrical system in the tank battery, when he came into contact with hydrogen sulfide gas causing him to fall on a nearby pipe which fractured his skull. The employee was found dead in the battery tank.
https://www.osha.gov/ords/imis/accidentsearch.accident_detail?id=70067.015

January 2015 OSHA Database: 1 Fatality

At approximately 1:45 p.m. on January 15, 2015, an employee was discovered on a tank battery catwalk with a respirator and gas monitor. The well site was known to be an active hydrogen sulfide environment. The employee was apparently gauging a tank. It is suspected that the employee was killed by over exposure to hydrogen sulfide or hydrocarbon vapors.
https://www.osha.gov/ords/imis/accidentsearch.accident_detail?id=70177.015

December 2014 OSHA Database: 1 Fatality

At 11:30 a.m. on December 1, 2014, Employee #1 was working on a heater treater, adjusting a valve to increase the flow when hazardous hydrogen sulfide gas escaped from the piping. He was overcome by the hydrogen sulfide gas and fell from an unspecified height, killing him.
https://www.osha.gov/ords/imis/accidentsearch.accident_detail?id=69267.015

October 2014 OSHA Database: 2 Fatalities

Attempted Rescue

At approximately 1:00 p.m. on October 22, 2014, Employee #1 was adjusting process piping at a waste water treatment plant that had been installed when he encountered a leak. Employee #1 exited Pump Room #2 to notify Water Utilities of the leak. Employee #1 returned to Pump Room #2 to try and close the main valve from the primary clarifier to isolate the flow from the leak. After entering Pump Room #2, Employee #1 was overcome with hydrogen sulfide and unable to exit the room. Upon hearing that Employee #1 had fallen, Employee #2 then entered the room to retrieve Employee #1. Both employees were unable to exit the room and were killed from over exposure to hydrogen sulfide.

https://www.osha.gov/ords/imis/accidentsearch.accident_detail?id=68644.015

July 2014 OSHA Database: 1 Fatality, 1 Injury (Hospitalization)

H_2S Explosion/Fire

At 9:45 a.m. on July 28, 2014, two employees were working as contract employees. With some coworkers, they were at a facility, preparing to cut piping on a "stick-water" storage tank. Stickwater is a slurry of water and fish matter. The contract workers were making a weld on piping attached to the lower part of the tank. This was hot work. They were preparing to cut piping on the top of the storage tank, when the top blew off. Employee #1 was thrown to the top of another tank. Employee #2 was thrown to the roof of a structure. Employee #1 who landed on the tank sustained massive head injury, multiple fractures, and he was killed. Employee #2 sustained multiple fractures and lacerations and he was hospitalized. It was later found that the air in the stickwater storage tank contained methane and hydrogen sulfide.

https://www.osha.gov/ords/imis/accidentsearch.accident_detail?id=65336.015

January 2014 OSHA Database: 1 Fatality

On January 23, 2014, Employee #1 fell approximately 10 feet to the ground from a tanker truck. He may have been overcome from hydrogen sulfide gas prior to his fall.

https://www.osha.gov/ords/imis/accidentsearch.accident_detail?id=202692471

October 2013 OSHA Database: 1 Fatality

At 9:30 a.m. on October 9, 2013, Employee #1 was removing a recycling line clamp. As the clamp was loosened, the pressure inside the recycling line blew the union apart, releasing oil and hydrogen sulfide gas. The employee was knocked to the ground and overcome by the hydrogen sulfide gas, asphyxiating and killing him.

https://www.osha.gov/ords/imis/accidentsearch.accident_detail?id=53287.015

July 2013 OSHA Database: 1 Fatality

At 9:00 a.m. on July 18, 2013, an employee climbed a caged ladder to open a hatch to gauge the water in a tank. The employee was overcome by hydrogen sulfide vapors and fell approximately 24 feet.

https://www.osha.gov/ords/imis/accidentsearch.accident_detail?id=48726.015

April 2013 OSHA Database: 1 Fatality, 1 Injury (Hospitalization)

Attempted Rescue

At 6:45 p.m. on April 11, 2013, an employee was cleaning the inside of an 8 foot diameter hydrogen sulfide sewer pipe using a high powered jet nozzle. He was working 125 feet inside the horizontal pipe without gas detection equipment. As the employee was cleaning the pipe with the high powered jet nozzle, the employee collapsed. A co-worker went to help him up and also collapsed. Employee #1 was killed and Employee #2 was hospitalized and survived.

https://www.osha.gov/ords/imis/accidentsearch.accident_detail?id=43504.015

January 2013 OSHA Database: 1 Fatality

At approximately 6:30 p.m. on January 17, 2013, Employee #1 was attempting to close a valve with a crescent wrench on a 3-in. pipe in the cellar of a pump jack. He was overcome by hydrogen sulfide being released from the pipe and is killed.

https://www.osha.gov/ords/imis/accidentsearch.accident_detail?id=38716.015

November 2012 OSHA Database: 1 Fatality

At 8:15 a.m. on November 6, 2012, an employee climbed a vertical ladder that was approximately 30 feet high to the top of a 20,000 gallon asphalt processing tank and was later found unresponsive on the upper working surface by a coworker. The medical examiner determined the cause of death to be related to fatal levels of hydrogen sulfide.

https://www.osha.gov/ords/imis/accidentsearch.accident_detail?id=33776.015

September 2012 OSHA Database: 1 Fatality

At 6:00 a.m. on September 30, 2012, an employee was working in a processing plant. Near the end of his shift, the employee was found unresponsive in the Desulfurization Area of the plant, lying near an open drain valve used to drain condensate from the sour gas line. As a result of the valve being left in the open position, the incident investigation surmised that the employee was in the process of manually draining the gas line when he became overcome by hydrogen sulfide (H_2S). The employee was asphyxiated as a result.

https://www.osha.gov/ords/imis/accidentsearch.accident_detail?id=30435.015

October 2011 California Fatality Assessment and Control Evaluation (CA/FACE) Report #11CA008) (2 Fatalities) (This appears to be a duplicate of the following 10/12/2011 report)

Attempted Rescue

Two laborers, 16- and 22-year-old brothers, died while cleaning the drainage system of an organic waste recycling facility. Both were exposed to excessive levels of hydrogen sulfide. The brothers were part of a three-man crew that was flushing out the drainage system. The 16-year-old was overcome by hydrogen sulfide and fell to the bottom of a 10-foot shaft. The 22-year-old collapsed at the bottom of the shaft after attempting to rescue his brother.

Contributing factors identified in this investigation were the high concentration of hydrogen sulfide in the shaft, failure to implement a confined space and hazard communication program, and the age of the youngest victim. The CA/FACE investigator determined that, in order to prevent exposures to hydrogen sulfide while cleaning out drainage systems, employers should:

- Develop, implement, and enforce a comprehensive confined space program.
- Develop, implement, and enforce a hazard communication program.
- Hire only employees aged 18 or older for any work tasks that may involve confined spaces.

October 2011 OSHA Database: 1 Fatalities, (1 Victim was hospitalized on life support)

Note: This incident was originally reported by OSHA to be 1 fatality – based on the above California FACE Report the second victim also perished due to H_2S poisoning.

At approximately 1:30 p.m. on October 12, 2011, Employee #1 and #2, farm workers, and a coworker were cleaning the storm drain system in preparation for the winter rains. Due to the amount of debris at the bottom of the 10-ft drain shaft, the water had backed up. They were told to use a 3-inch pressure hose from the water truck to force the debris down the system to a drainage pond. Employee #1 sat on the edge of the 24-inch concrete pipe shaft and tried to break up the clog. It was hard to handle the hose, so the coworkers picked up half of an extension ladder that was leaning against a fence and put it inside the shaft, so that he could brace his feet against the rungs. Employee #2 saw Employee #1 standing on the ladder in the hole, at about shoulder level, with his head above grade. He said that he didn't feel well and was getting sleepy. He then descended the ladder, to the bottom of the hole. It is unclear if he stepped down the ladder or if he was losing consciousness and slid down. Employee #2, who was standing outside of the hole, saw what had happened and attempted to go into the hole to rescue Employee #1. The coworker tried to prevent Employee #2 from going into the hole but could not. Employee #2 was also overcome and collapsed to the bottom of the shaft. Emergency medical services were called. When emergency response teams arrived, they did a direct reading of the shaft and the instrument topped out at 200 ppm of hydrogen sulfide. Employee #1 and #2 were transported to Kern Medical Center, where Employee #1 was pronounced dead, and Employee #2 was put on life support.

https://www.osha.gov/ords/imis/accidentsearch.accident_detail?id=201495637

October 2011 OSHA Database: 1 Fatality, 1 Injury (Treated and released)

Attempted Rescue

On October 16, 2011, an employee was working as the owner of Esther's Restaurant in Miami, Florida. He entered a sewer manhole to replace some pumps. Upon entry into the manhole, he lost consciousness. When a coworker entered the space, he fell ill and had to come out. No testing or breathing precautions were taken by either the restaurant owner or the coworker. The restaurant owner was taken to a local hospital, where he was pronounced dead. The coworker who had attempted a rescue was taken to a hospital, where he was treated and released to go home. Only the restaurant's owner was listed on the injury line. There was no mention of the coworker. According to the information gathered from the local authorities, the composition of the ambient air within the manhole was measured at 200 parts per million of hydrogen sulfide and 1200 parts per million of carbon monoxide. The narrative says that at approximately 12:27 p.m. on October 26,

2011, OSHA's Ft Lauderdale, Florida Area Office received a call from Miami-Dade Fire Rescue (the fire department) to report the fatality. The date might have been entered incorrectly.

https://www.osha.gov/ords/imis/accidentsearch.accident_detail?id=200678449

September 2011 OSHA Database: 2 Fatalities

On September 21, 2011 Employee #1 and Employee #2 of Stevens Electric & Pump Service, Inc., were found dead inside a sewer tank, a confined space. They died from exposure to hydrogen sulfide. The space had been improperly isolated, no air monitoring had been performed, no air supplied to the space and no retrieval system used by the person entering into the confined space.

https://www.osha.gov/ords/imis/accidentsearch.accident_detail?id=202220836

March 2011 OSHA Database: 1 Fatality, 1 Injury (Minor)

At approximately 3:00 p.m. on March 6, 2011, ten workers were performing maintenance on a 180-ft process tower. Employee #1 and a coworker were doing site preparation to install a 20-inch nozzle on the tower at the 100-ft level. Five of the ten workers were at the very top of the tower preparing to do confined space entry for internal tower work. Three workers were at the ground level doing structural work. The personal hydrogen sulfide (H_2S) meters on Employee #1 and his coworker started to alarm at the same time. The coworker called out on his radio, alerting the other workers, including a ground level supervisor, of the encounter with the H_2S. Immediately, all workers on the tower began to descend from their respective levels. Employee #1 and his coworker began descending from the 100-ft level and encountered an increasingly dangerous atmosphere. As those workers at the very top began to descend, their personal meters began alarming. The ground level superintendent suspected that the H_2S was at a higher concentration near the 100-ft level and called out on the radio for those workers that were in the process of descending from the very top to go back up. Employee #1 and his coworker began to lose consciousness and fell from their respective locations. Employee #1 fell down to the deck near the 100-ft level. His coworker slid down the ladder cage that connected the deck near the 100-ft level to the deck below. The coworker slid down into a more breathable atmosphere, sustaining bruising but survived with minor injuries. Employee #1 passed out from the H_2S gas and suffered blunt trauma to his head, resulting in his death.

https://www.osha.gov/ords/imis/accidentsearch.accident_detail?id=200556389

February 2011 OSHA Database: 0 Fatalities 0 Recordable injuries

H_2S Released - Mixing Chemicals

February 25, 1011–Employee #1 mistakenly mixed two incompatible substances together, which created a release of hydrogen sulfide. Employee #1 was performing a routine task, did not read the labels of the chemical containers prior to performing a chemical transfer. Employee #1 was not hospitalized.

https://www.osha.gov/ords/imis/accidentsearch.accident_detail?id=201128360

February 2010 OSHA Database: 1 Fatality

At approximately 2:30 p.m. on February 17, 2010, Employee #1 was working on a hog farm. He was in the utility hallway of a hog barn. The hog barn consisted of two confinement rooms (east and west) separated by the utility hallway. Employee #1's responsibilities consisted of checking the health of the hogs, which included providing them with food and water and administering antibiotic shots. Chores for each building took approximately 30 to 40 minutes, or approximately 15 minutes for a half-full building. At the time of the incident, a coworker was pumping the manure holding tank from beneath the west confinement room. The coworker stated that he saw Employee #1 arrive on site, but he didn't know when Employee #1 entered the building. Employee #1's coworker had started pumping around 8:45 a.m. that morning and was almost finished when the incident occurred. Manure was pumped from the south side of the building into a tank that was pulled by a tractor to an offsite concrete pit and unloaded. According to two other coworkers, one of the two manure pit fans was ordinarily disconnected while the pumping took place, because the pump shaft was placed into the fan hole. The coworker pumping manure did not notice anything unusual during the process. Employee #1 was working alone and was found unconscious at the north end of the utility hallway that separates the west and east confinement rooms. According to the final coroner's report, the immediate cause of death was acute respiratory failure due to hydrogen sulfide poisoning. When the fire department responded to the incident at 3:17 p.m., air measurements were collected inside the building using a four-gas meter. Hydrogen sulfide was measured at 42 to 44 ppm near the north service door, and carbon monoxide was measured at 30 ppm in the east hog confinement area. According to the sheriff's report, the north and south service doors had been open for an unknown amount of time, which most likely ventilated the hallway prior to the fire department's air measurements.

https://www.osha.gov/ords/imis/accidentsearch.accident_detail?id=200514990

January 2010 OSHA Database: 1 Fatality, 1 Injury requiring hospitalization, 5 Injuries not requiring hospitalization.

On January 9, 2010, at least four Employees of Enbridge G & P (East Texas) working under the supervision of the plant manager were involved in relighting a waste heat boiler/reaction furnace. The process to restart the waste heat boiler/reaction furnace required the employees to stand close the sight glass valve, remove the sight glass, slide the igniter into the valve, open valve, slide igniter into the burner, ignite, pull igniter partially out, close valve and replace sight glass. However, it was discovered that the sight glass valve had seized and needed to be replaced. Without completing a safe work permit, locking and tagging out the system, or purging it of any toxic or flammable substances, the employees began replacing the valve when the hydrogen sulfide release occurred. Employee #1 died from hydrogen sulfide poisoning, Employee #2 was hospitalized from hydrogen sulfide poisoning, and Employees #3, #4, #5, #6, and #7 suffered symptoms of hydrogen sulfide poisoning but did not require hospitalization.

https://www.osha.gov/ords/imis/accidentsearch.accident_detail?id=200784957

September 2009 OSHA Database: 1 Fatality, I Injury (Hospitalized), 3 Injuries not requiring hospitalization.

Attempted Rescue

On September 10, 2009, five employees were working on the site of a nonproductive natural gas well, for Chipco, LLC, a gas well service company. The well was being plugged and abandoned, and the materials were to be salvaged. Water was being poured into the well. The water circulated, rose to the top, and flowed into a trench and pit which had been dug for this purpose. The pit liner started to slide into the pit. Employee #1 ran over to pull it back up and secure to the side of the pit. A pocket of hydrogen sulfide released from the top of the pipe. Employee #1 was asphyxiated and died. A coworker working nearby was asphyxiated but was not killed. Three other employees ran to the site but were unable to assist because they became incapacitated by the gas. Employee #2 was hospitalized for asphyxiation. Employees #3, #4 and #5 were not hospitalized. The owner of Chipco, LLC was aware of the presence of hydrogen sulfide at this particular well site. He had completed a walk thru inspection of the site with the previous owner, Northcoast Energy, prior to start of work. He had been informed by the representative of Northcoast Energy that this well was being capped due to it containing hydrogen sulfide. He also observed corrosion around the well head which is an indicator of hydrogen sulfide. He did not provide any type of personal or area air monitoring of the site on the day of the incident. The employees were not provided with escape type respirators. Signs were not posted on the site to indicate the presence of hydrogen sulfide gas.

https://www.osha.gov/ords/imis/accidentsearch.accident_detail?id=200758878

June 2009 OSHA Database: 3 Fatalities

Attempted Rescue

On June 29, 2009, Employee #1, who worked for S. Dahan Piping and Heating Corp., entered a dry well (confined space) through a 3-ft-diameter opening. As he descended an 18-ft ladder, he was overcome by hydrogen sulfide. Employee #2, who worked for the same company, attempted to rescue Employee #1, and he too was overcome by the toxic gas. Employee #3, employed by Regal Recycling Co., Inc., also entered the confined space to attempt rescue of the first two workers, and was also asphyxiated. All three workers died at the scene.

https://www.osha.gov/ords/imis/accidentsearch.accident_detail?id=200831469

April 2009 OSHA Database: 1 Fatality

On or about April 14, 2009, Employee #1, a truck driver, pulled his tanker truck containing a load of naphthenic waste caustic from Citgo refinery in Lake Charles, LA, adjacent to a processing tank NV-1, and connected it to discharge his load to the process tank. He was observed opening the dome lid to the tanker. Evidence suggests he may have been exposed to hydrogen sulfide while opening the dome lid, became light-headed, and may have fallen either on the top of the tanker, or as he attempted to descend the ladder from the tanker. He apparently then sat in a chair about 50 feet from the process vessel, and then toppled from the chair, unresponsive. There was conflicting information regarding whether he was administered oxygen. A coworker called 911, and another performed CPR. Employee #1 was declared dead on arrival at the hospital.

https://www.osha.gov/ords/imis/accidentsearch.accident_detail?id=201924859

January 2009 OSHA Database: 1 Fatality, 3 Injuries (Hospitalization)

On January 11, 2009, Employees #1, #2, #3, and #4 were replacing a vent line with a vapor removal line to remove hydrogen sulfide from a tank when they were overcome by the gas fumes. Employee #1 was rendered unconscious and died, while Employees #2, #3, and #4 were taken by air to the nearby regional hospital, where they were hospitalized and treated for asphyxia.

https://www.osha.gov/ords/imis/accidentsearch.accident_detail?id=200812964

September 2008 OSHA Database: 0 Fatalities, 1 Injury (Hospitalization)

At approximately 7:30 a.m. on September 24, 2008, Employee #1, the Night Shift Supervisor, was discovered unresponsive at the bottom of the tank of a tanker truck. Hydrogen sulfide gas levels of 60 parts per million to 200 parts per million were detected in the tank by the Fire Department. Employee #1 was working at a nonunion, recycling operation where they collect used cooking oils and greases from oil water separators and other reservoirs at restaurants and food production facilities and process these waste materials for resale as various animal feed product additives. Employee #1 was cleaning the inside of one of the tanker trailers that had been used to collect waste cooking oils and greases mixed with water. The tanker was not draining properly and spraying water into the tank from the open hatch was not effective so Employee #1 climbed through the 20-in. wide access hatch at the top of the 40- ft long, 6,500 gallon tanker. Employee #1 was asphyxiated and was hospitalized as an inpatient for five days.

https://www.osha.gov/ords/imis/accidentsearch.accident_detail?id=202584629

July 2008 OSHA Database: 2 Fatalities

Attempted Rescue

On July 21, 2008, Employee #1 entered a sewer lift station to check for leaks in the line. Employee #1 was overcome by hydrogen sulfide gas. A second employee entered the station to retrieve Employee #1 and also was overcome by the gas. Both employees died from overexposure to hydrogen sulfide gas.

https://www.osha.gov/ords/imis/accidentsearch.accident_detail?id=200332104

July 2008 OSHA Database: 0 Fatalities, 2 Injuries (1 Hospitalized, 1 Treated and released)

Attempted Rescue

7/8/2008: The pipe laying crew assigned to the jobsite was responsible for replacing a 45-ft section of sewage pipe to an already existing sewage system. At approximately 8:15 a.m. Employee #1, the foreman, directed Employee #2, a pipe layer, into a 10-ft deep manhole to secure a leaking air plug that had jarred loose from installing the trench box and shields. Employee #2 entered the manhole to secure the air plug, became short of breath, and ascended to the top of the manhole to take a short break to wipe his face. Employee #2 descended the manhole ladder to secure the air plug, was overcome by hydrogen sulfide and fell approximately 5-ft to the bottom of the bottom of the manhole. The manhole attendant screamed for help and Employee #1 descended the manhole ladder in an attempt of rescue the unconscious Employee #2. Employee #1 was then overcome by hydrogen sulfide, loss consciousness, and fell to the bottom of the manhole. At approximately 8:25 a.m. the excavator operator called Emergency Services and prevented any employee from entering the manhole. The excavator operator responded by directing air from the compressor into the manhole enough for both employees to regain consciousness. A strap was fixed to the excavator bucket and lowered into the manhole to help the employees escape. Employee #1 used the strap and manhole ladder to extricate himself from the manhole. Employee #1 had also used the strap and manhole ladder, but fell back into the manhole once before escaping. At approximately 8:30 a.m. the Fire and Rescue Squad arrived to administer medical treatment until the ambulances arrived. Both men were sent via ambulance to the hospital. Employee #1 was admitted to the hospital for two days for hydrogen sulfide poisoning and head contusions. Employee #2 was seen in the trauma center for hydrogen sulfide poisoning, neck, head and knee contusions and was released that day. No confined space permit, atmospheric testing, or monitoring was completed prior to employees entering manhole. The injury line in the original report does not list Employee #2.

https://www.osha.gov/ords/imis/accidentsearch.accident_detail?id=200925485

June 2008 OSHA Database: 1 Fatality, 1 Injury (Hospitalized) 2 Injuries (Treated and released)

Mixing chemicals

At approximately 10:30 a.m. on June 21, 2008, an employee was processing a 5,000 gallon capacity tank containing nonhazardous waste water plus a precipitating agent called Sodium Hydrogen Sulfide. The employee became exposed to a major release of Hydrogen Sulfide gas which caused him to lose consciousness and collapse onto a catwalk located approximately 12 to 15 feet above the floor of the building. Three coworkers who happened to be in this building due to working on a renovation project heard coughing sounds. They made an attempt to ascend the stairs in order to check on the employee, but they encountered a strong rotten egg odor in the air shortly after seeing the employee lying in an apparently unconscious state on his back on the floor of the catwalk. Coworker #1 screamed out:

"It's Hydrogen Sulfide! We need to get out of here now!" All coworkers quickly evacuated the building and called the emergency medical services. They alerted the emergency medical services Operator about the presence of Hydrogen Sulfide. Coworker #1 pulled a gas meter out of his vehicle and observed a reading of 24 ppm of Hydrogen Sulfide approximately 10 to 20 feet outside of the back building containing the employee. Several of the responding Middletown Fire Department workers also smelled a rotten eggs odor, some as much as 80 to 100 feet away from the back building. They donned SCBA's before entering this building and transporting the employee toward a waiting ambulance and then on toward Atrium Medical Center in Middletown, Ohio. Two Middletown Fire Department workers were checked out by this medical facility because they had experienced abnormal symptoms such as chest tightness, throat numbness and coughing up of excess sputum. On June 24, 2008, OSHA detected the following lingering levels of Hydrogen Sulfide gas: 0 ppm in the front building, 1 ppm on the ground level of the back building, 2 to 3 ppm on the catwalk level, and 42 ppm approximately 1 foot below the opening in the upper surface of Tank Number 204. The employee died of overexposure to Hydrogen sulfide.

https://www.osha.gov/ords/imis/accidentsearch.accident_detail?id=201955549

December 2007 OSHA Database: 2 Fatalities

Attempted Rescue

At 5:00 a.m. on December 2, 2007, Employee #1, a tank cleaning technician for Grand Isle Shipyard, was spraying wash water in the port side holding tank of the utility ship "Terry Bordelon." Employee #1 also had an attendant perched above him at the top of the entrance to the tank, which is the deck of the boat. Employee #1, in the portside tank, signaled to his attendant that he needed a rag to wipe the glass on his full facepiece, airline respirator. The attendant got a rag and tossed it down to Employee #1, who began to climb up the tank ladder to exit the tank. When Employee #1 reached the top rung, he fell back down into the tank, displacing his respirator and exposing him to the tank atmosphere. His attendant called for help from the site supervisor, and a rescue tripod was acquired a short distance away from the entrance to the port side tank. The other starboard side cleaning technician was summoned to enter the port side tank for rescue. While the port side attendant and site supervisor were obtaining the tripod, the other starboard side attendant (Employee #2), working at the starboard side tank, on the adjacent job, entered the portside tank to reach Employee #1 lying on the tank floor. The cleaning tech from the starboard side tank entered into the port side tank to attach the tripod line to Employee #1's harness. When he went down into the tank, he observed the starboard side attendant (Employee #2) laying unconscious without an airline respirator. The injured cleaning tech was immediately retrieved. Emergency Medical Services had cut the harness from Employee #1, rendering it unusable for Employee #2. Another harness had to be obtained from the ships crew. By the time the starboard side attendant (Employee #2) had been retrieved, he was unconscious and could not be revived and shortly died. Employee #1 died a few days later from the injuries he sustained from inhaling hydrogen sulfide gas.

https://www.osha.gov/ords/imis/accidentsearch.accident_detail?id=200555076

November 2007 OSHA Database (Two separate reports): 3 Fatalities

Two brothers and Two others were killed in this incident (Please See This Report and the Following Report)

At approximately 5:30 p.m. on November 1, 2007, three employees were replacing a sump pump in an underground lift station which collected drainage and leached water from a construction debris landfill. Employee #1 entered the station and commenced work at which time he was overcome by hydrogen sulfide gas. Employee #2 noted that Employee #1 needed assistance and entered the station to assist, at which time Employee #2 was overcome by the hydrogen sulfide gas. Employee #3 noted that the first two Employees were in trouble and also entered the station to assist. Employee #3 was also overcome by hydrogen sulfide gas. Rescue services arrived on site and performed air quality monitoring which revealed that all three victims were exposed to concentrations of up to 200 PPM of hydrogen sulfide gas. Body retrievals were initiated at that point. The lift station was determined to be a permit-required confined space. The company had not evaluated the lift station to determine that it was a permit-required space and had not developed and implemented a written permit space program. All three employees were killed.

https://www.osha.gov/ords/imis/accidentsearch.accident_detail?id=200340743

November 2007 OSHA Database: 1 Fatality

This Report Is for the Other Person Killed in the Same Incident.

At approximately 5:30 p.m. on November 1, 2007, Employee #1, of Midwest Boring & Underground, Inc., was assisting another company with replacing a sump pump in an underground lift station which collected draining and leached water from a construction debris landfill. Three employees of the other company entered the lift station and were overcome by exposure to hydrogen sulfide gas. Employee #1 entered the lift station to assist or rescue the three workers when he, too, was overcome. Employee #1 was killed. Rescue services arrived at the scene and performed air quality monitoring, which revealed that all four workers were exposed to concentrations of up to 200 ppm of hydrogen sulfide gas. The lift station was determined to be a permit-required confined space. The other company, which was the host employer, had not evaluated the lift station to determine whether it was a permit-required space. Neither employee had developed and implemented a written permit space program.

https://www.osha.gov/ords/imis/accidentsearch.accident_detail?id=200340750

October 2007 OSHA Database: 1 Fatality

At approximately 9:00 a.m. on October 22, 2007, two employees parked a vacuum truck along the roadside near the Number 10 Reformer in a Refinery. The purpose was to vacuum liquids from the Y-3 Acid Sump and Skid. One employee began collecting liquid from the Y-3 acid sump through a hose into a vacuum truck parked approximately 40 ft away from the Y-3 sump. Employee #1 was standing at the rear of the vacuum truck operating the control valves. Several minutes into the vacuum operation, Employee #1 yelled to his coworker that he smelled a very strong odor. The coworker advised Employee #1 to move away from the vacuum truck moving beyond the other side of the vehicle. Employee #1 stepped out from behind the vacuum truck along the passenger side of the vacuum truck. A "gooseneck" vent pointing in a downward angle at the front portion of passenger side of the truck was venting copious amount of concentrated vapors consisting of Hydrogen Sulfide and possibly Methyl Mercaptan. Employee #1 inhaled the vapors and staggered to the other side of the plant road where he attempted to sit down but fell over according to witnesses. The coworker ran over to the vacuum truck to shut off the pump and proceeded around the front of the vehicle. At this time, Employee #1 had walked back to the vacuum truck and placed himself in the path of the vent again. Employee #1 collapsed. Two refinery employees, outside operators, that witnessed Employee #1 stagger across the street had retreated to a nearby control room to don fresh air packs. These outside operators removed Employee #1 from the location along side the passenger side of the vacuum truck. The outside operators attempted to revive Employee #1 who died due to the exposure.

https://www.osha.gov/ords/imis/accidentsearch.accident_detail?id=200555043

August 2007 OSHA Database 0 Fatalities, 1 Injury (Hospitalization)

H_2S explosion burned 85% and Multiple Broken Bones and Internal Injuries

August 20, 2007 Employee #1, an independent contractor, working by contract with an oil company, was to check the tank raw oil levels on the three tanks daily at this site. One of the tanks, a 210-barrel stock tank had recently been cleaned by another contracting company. The access panel at the lower section was not put back on. Employee #1 had not been told to put the panel back on the tank. Hydrogen sulfide inside the tank (a condition that occurs in this industry at times) exploded. The explosion propelled Employee #1 into his parked truck and onto the field, for a total of approximately 40 yards. Employee #1 suffered multiple broken bones, internal injures and burns over 85 percent of his body. The explosion occurred from Employee #1 using an electric impact wrench to remove bolts from the access panel on the tank. The proper procedure to secure the panel would include ventilating the tank by mechanical fans and then using non spark producing wrenches to open or close the access panel. The reason for the NO INSPECTION on this accident was because the lack of an Employer-Employee relationship. Employee #1 was an Independent Contractor. A copy of the contract between Employee #1 and the oil company was provided. Interviewed company people, Employee #1's family member, other Independent Contractors working for the oil company were in agreement. The Attorney for Employee #1 agreed that Employee #1 was an Independent Contractor. Employee #1 was working by himself, checking on various oil wells daily. An Employer-Employee relationship was not present, thus there is no jurisdiction over this accident.

https://www.osha.gov/ords/imis/accidentsearch.accident_detail?id=200841575

August 2007 OSHA Database: 1 Fatality

On August 10, 2007, Employee #1 entered a 10.5 ft deep manhole to remove three line plugs and activate the sewer system when he was overcome, probably by hydrogen sulfide. Employee #1 died of asphyxia. The manhole had not been previously entered and was not being monitored for toxicity, oxygen level, or explosive levels. No tripod was in place for emergency retrieval of personnel in the manhole. At the time this this report was written, the probable cause of death had been identified as hydrogen sulfide poisoning, according to the Pima County Forensic Science Department

https://www.osha.gov/ords/imis/accidentsearch.accident_detail?id=201992997

March 2007 OSHA Database: 0 Fatalities, 4 Injuries (Examined and released at hospital)

On March 14, 2007, Employees #1 through #6 were working at a facility where ophthalmic goods such as ophthalmic frames, lenses, and sunglass lenses were manufactured. A high-index plastic lens with a refractive index of 1.74 was being ground without a water wash. The plastic overheated and emitted a small amount of hydrogen sulfide gas, which the employees inhaled. The six employees displayed a physical reaction to the strong odor. Employees #1 and #2 were examined at the scene and determined not to need further treatment. Employees #3 through #6 were taken to the hospital, where they were examined and released. (The original form did not list Employees #3 through #6 on an injury line.)

https://www.osha.gov/ords/imis/accidentsearch.accident_detail?id=200073120

January 2007 OSHA Database: 1 Fatality

On January 20, 2007, Employee #1 was working in the Hydrofluoric Acid, (HF), and Alkylation Unit in charge of maintaining the neutralization pits. He was in charge of draining a constant boiling mixture, (CBM's), which contained HF acid and water out of the polymer surge drum into a neutralization pit. The CBM's, low in pH, and were sent to the South neutralization pit to neutralize spent potassium hydroxide that was contained in the open-surfaced pit. Potassium hydroxide, (KOH), is highly basic and CBM's are drained into the South pit to lower or neutralize the pH. The draining of CBM's is accomplished by pressurizing the polymer surge drum with a constant supply of natural gas. The vessel is pressurized to 25 to 30 pounds and then a drain valve is manually opened and CBM's are forced out into the South neutralization pit. The operator visually watches the CBM's come out of a discharge pipe into the South neutralization pit. When the operator sees a color change, indicating that the CBM's have been evacuated from the vessel and polymer is now entering the South pit, the drain valve is then closed. The pH is checked manually with the use of litmus paper. When the solution in the South pit has reached a pH range of 7 to 9, the contents are then sent to the North pit for further treatment. Once the mixture is fully neutralized it is sent to waste water. On the January 19th, 2007, the operator had drained the CBM's out of the polymer surge drum and performed an unusually long draw. The draining of CBM's, which normally takes seconds, had taken three minutes. Depending on the position of the drain valve this could have introduced a large volume of polymer into the South pit. Polymer or acid soluble oils, (ASO), have a low pH and contain sulfur compounds. Later that day, a large volume of spent KOH was sent to the South pit, approximately 940 gallons. This would have been performed to further regulate pH. On the January 20th, 2007, the liquid level in the South pit was close to the top edge. At some point the drain valve on the South pit was turned in the open position. There is only one reason the drain on the South pit would have been opened; to lower the level in the pit because the contents were going to flow over the top of the pit. The drain on the South pit is normally closed, due to the fact that the contents have to be sent to the North pit for further treatment. On January 20, 2007, Employee #1 opened the drain on the South pit and the head pressure from the South pit released vapors. The vapors traveled up to Employee #1 who lost consciousness. No respiratory protection was worn by Employee #1, and he was wearing an H_2S dosimeter. It is undetermined at this time if the dosimeter was working properly. When coworkers found Employee #1, they stated that the H_2S monitor was alarming. The highest reading from the dosimeter was 95 ppm. The wind was blowing from the South, but steam traps showed that the wind was also swirling near the pit. Employee #1 died from acute hydrogen sulfide intoxication.

https://www.osha.gov/ords/imis/accidentsearch.accident_detail?id=200841302

September 2006 OSHA Database: 1 Fatality

At approximately 11:00 a.m. on September 5, 2006, Employee #1, a sewer worker, entered a sewer to manually clean and vacuum. Two other workers stayed at street level. No atmospheric readings were taken prior to the worker's entry into this combination sanitary and industrial 60-inch sewer that was approximately 35 feet below street level. Employee #1 entered the sewer without any respiratory protection and was not tied off for retrieval or emergency rescue. He worked in the sewer for approximately one hour and during this time the two other workers, both of whom were at street level, observed that the odor coming from the sewer was becoming stronger. Employee #1 came back to street level for approximately 10 minutes, complaining of difficulty breathing. Against the warnings of the other two workers, he went back into the sewer, again without any atmospheric readings being taken and no personal protective or rescue equipment. A few moments later, the attendant was not able to communicate with Employee #1 and called emergency services. Rescue crews arrived, retrieved him from the sewer, and transported him to the hospital where he died less than 24 hours later. Readings obtained by the rescue crew, prior to their entry, indicated high levels of hydrogen sulfide. Readings obtained from the Northeast Ohio Sewer department, taken shortly after the accident, showed high levels of ammonia.

https://www.osha.gov/ords/imis/accidentsearch.accident_detail?id=200102515

June 2006 OSHA Database: 1 Fatality

Attempted Rescue

On June 5, 2006, Charco Maintenance & Welding was hired by the facility owner, Petro Operating Company, to do maintenance and repair work during a shutdown of the facility. The four-man crew was replacing some valves that were attached to a 10,000-gallon capacity surge tank. The surge tank contained saltwater that was mixed with hydrogen sulfide, carbon dioxide, and hydrocarbons. The contents of the surge tank were lowered to a level below, where the valves were attached. Employees #1 and #3 climbed the fixed ladder of the tank to one of the elevated platforms to remove one of the valves. They used a MiniCheck hydrogen sulfide detector, which was manufactured by Del Mar Scientific that changes color when hydrogen sulfide is detected. They used the detector, when the valve was initially loose from its connection, but the detector did not indicate any hydrogen sulfide. During the removal of the valve, hydrogen sulfide was released into their immediate work area. It was estimated that the airborne concentration of hydrogen sulfide was approximately 500 parts per million. Employees #1 and #3 lost consciousness from the exposure. When Employee #3 regained consciousness, he climbed down from the platform. Employee #4 climbed the ladder to help Employee #1. He removed his SCBA mask to provide clean air to Employee #1, but was almost over come, and had to redon his mask. Employee #1 was lowered to the ground, but did not regain consciousness, and later died at the scene from his exposure. Employees #3 and #4 were not hospitalized for their exposure. Employee #2 was not exposed. It was determined that the employees had not been properly trained on the lock-out/tag-out program, handling the potential exposures to air contaminants, or rescue efforts.

https://www.osha.gov/ords/imis/accidentsearch.accident_detail?id=201353117

May 2006 OSHA Database: 2 Fatalities

Attempted Rescue

On May 19, 2006, Employee #1 and Employee #2 were working at a private residence. Employees #1 and #2 were irrigating a well to determine why the well was not producing the required amount of pressure to irrigate the homeowner's yard. While they were working, Employee #2 walked up to a 8-ft by 8-ft concrete pit, where the well was located, and noticed Employee #1 lying face down in the standing water at the bottom of the pit. Employee #2 ran to his truck to retrieve a cell phone, threw his phone to a coworker standing nearby, and asked him to call emergency personnel. Meanwhile, Employee #2 climbed down a ladder to enter the pit to assist Employee #1.

As soon as Employee #2 head was level with the ground he was also overcome and fell on top of Employee #1 in the bottom of the pit. Employee #1 and Employee #2 were killed by exposure to by hydrogen sulfide gas.

https://www.osha.gov/ords/imis/accidentsearch.accident_detail?id=200271567

March 2006 OSHA Database: 2 Fatalities

Attempted Rescue

On March 21, 2006, Employees #1 and #2 were pumping the water out of the compartments of a casino barge section that was capsized during hurricane Katrina, in preparation to salvage the barge. Employee entered the aft rake compartment of the barge and apparently was asphyxiated by fumes, and he drowned. Employee #2 was looking for the missing employee, when he spotted his hard hat and personal floatation vest laying next to the opening to the aft rake compartment. Employee #2 went to the area and looked in. He saw Employee #1 floating face down in the compartment. Employee #2 yelled for someone to call for ambulance and subsequently entered the same compartment. He was also overcome by the fumes and drowned. The supervisor stated that there was a strong smell of rotten eggs, hydrogen sulfide, coming from the compartment, when he got to the opening to the compartment. The supervisor reported this to the first emergency services team and requested to arrive on scene. The emergency services confined space team was called to the scene and entered the compartment after venting the space. Both employees were recovered, and they were dead.

https://www.osha.gov/ords/imis/accidentsearch.accident_detail?id=201362605

February 2006 OSHA Database: 1 Fatality

Attempted Rescue

At approximately 9:00 a.m. on February 15, 2006, Employee #1 and Employee #2 were unclogging a plugged steam ejector and became unconscious. Apparently, an organo-sulfur compound decomposed, liberating hydrogen sulfide and carbon monoxide when it became exposed to temperatures in the vicinity of 160-degrees Celsius. No local monitoring for hydrogen sulfide, carbon monoxide, oxygen or LEL was performed prior to the start of the maintenance. Employee #1 passed out shortly after the line was broken by the removal of a flange. Employee #2 contacted the Emergency Response Team indicating that it appeared Employee #1 was experiencing a heart attack. Employee #2 began to lift up and move Employee #1, but he, too passed out. Emergency Response Team members arrived.

CPR was performed. Both were rushed to the hospital. Employee #2 gradually recovered and was released from the hospital on February 19, 2006. Employee #1 never came out of his coma and died on February 18, 2006.

https://www.osha.gov/ords/imis/accidentsearch.accident_detail?id=201954500

October 2005 OSHA Database: 1 Fatality

October 22, 2005: Employee #1 opened a cover on the top of an oil storage tank in order to gage the amount of water in the tank and was overcome by hydrogen sulfide gas. Employee #1 was killed.

https://www.osha.gov/ords/imis/accidentsearch.accident_detail?id=200603157

September 2005 Report by *Acute Communicable Disease Control (ACDC) 3 Fatalities, 19 Possible Injuries

On September 2, 2005, ACDC was notified of three deaths occurring in the propeller room of a cruise ship docked in the Port of Los Angeles. Initial reports suggested that the deaths were related to hydrogen sulfide poisoning occurring while crew-members were attempting to fix a pipe in the propeller room on the ship. Hazmat obtained an environmental air sample at 11:00 am on September 3, 2005—5 hours after the incident in the propeller room—measuring 100 ppm of H_2S.

August 2005 OSHA Database 0 Fatalities, 4 Injuries (Hospitalized)

On August 9, 2005, five employees were working in a sewage treatment plant to unclog a sludge line. As they were working to free the line, they were overcome by hydrogen sulfide gas. Employees #1, #2, #4, and #5 were hospitalized and treated for asphyxia, while Employee #3 received no serious injuries.

https://www.osha.gov/ords/imis/accidentsearch.accident_detail?id=201857844

March 2004 OSHA Database: 0 Fatalities, 3 Injuries (Hospitalized)

Attempted Rescue

At approximately 12:15 p.m. on March 9, 2004, three employees were pumping a muriatic and water solution through a two-inch line. The line went into a 500-gallon portable storage tank. Employee #1 was overcome by hydrogen sulfide fumes and passed out. Employees #2 and #3 both attempted to rescue Employee #1 but were also overcome by the fumes. All three employees were hospitalized and released the next day.

https://www.osha.gov/ords/imis/accidentsearch.accident_detail?id=200553287

February 2004 OSHA Database: 0 Fatalities, 1 Injury (Hospitalization)

Employee #1 was performing a coil turning job on an oil well when he was overcome by hydrogen sulfide gas. He was hospitalized overnight for observation and was released the next day.

https://www.osha.gov/ords/imis/accidentsearch.accident_detail?id=200980399

November 2003 OSHA Database: 2 Fatalities, 1 Injury (Hospitalized)

On November 11, 2003, Employees #1 and #2, along with one additional coworker, were working in a 36-inch sewer conduit to pressure wash and remove debris. Entry into the sewer system was conducted through a 24-inch-diameter manway with a 17-foot vertical descent by ladder. While in the sewer main, Employees #1 and #2 and the coworker were overcome by toxic gas. Employees #1 and #2 died from exposure to hydrogen sulfide. The coworker was exposed, but was revived. The required preparatory steps, for this type of work, were not followed with any diversion of the flow of raw sewage, atmospheric testing conducted, and mechanical ventilation provided. Additionally, the three workers were not provided with personal gas monitors, respiratory protective equipment, and rescue equipment.

https://www.osha.gov/ords/imis/accidentsearch.accident_detail?id=200211399

October 2003 OSHA Database: 1 Fatality, 3 Injuries (Hospitalized)

Attempted Rescue

On October 10, 2003, Employee #1 climbed up to a platform approximately 6 ft above the floor to re-tighten the gasket packing on the hydrolyzer machine. The hydrolyzer cooks down poultry feathers to produce a protein ingredient for animal food at this rendering plant. The process of decomposing feathers was known by the plant to create hydrogen sulfide gas inside and around this hydrolyzer. The employee was attempting to tighten the packing while the hydrolyzer continued to run and spray out steam and product at the feed end of the machine around the center shaft. Employee #2, a supervisor, was watching from the floor level below when Employee #1 stood up and leaned against a guardrail. Employee #2 noticed that Employee #1 was bleeding from his nose. Employee #1 then collapsed to the catwalk platform. Employee #2 called for help and climbed up on the platform to assist Employee #1, while other coworkers responded. Employee #2 collapsed after Employee #1 was removed from the catwalk platform. Both employees were removed from the area and Employee #1 was given CPR until paramedics arrived. The two paramedics continued to treat Employee #1, but never revived him and he was pronounced dead at the scene. The two paramedics, Employee #3, and 4 also were overcome by hydrogen sulfide, noting a rotten egg smell while treating Employee #1 in the back of their ambulance. Two more ambulances were dispatched, one treating and transporting Employee #2, and one assisting and transporting the two paramedics.. Employee #2 and the two paramedics were admitted to the hospital for exposure to hydrogen sulfide.

https://www.osha.gov/ords/imis/accidentsearch.accident_detail?id=200262178

September 2003 OSHA Database: 2 Fatalities, 1 Injury (Hospitalized and released)

Attempted Rescue

On September 12, 2003, Employee #1 and 2, a two man crew from a pump service company had worked most of the day to repair a pump located near the bottom of 21-ft deep domestic sewage lift station. They had been supported by a two to three man crew from the Public Service District that owned the lift station and the sewage collection system. The atmosphere in lift station had been continuously monitored with a multi- meter for explosive gasses, oxygen content and hydrogen sulfide. They had completed the work, tested the pump and unblocked inlet pipes to permit sewage to flow into the lift station. They packed the multi-meter, tools and equipment on to the service truck in preparation of leaving the site. Employee #1 was retrieving the last piece of equipment, a two part extension ladder used to access the bottom of the lift station. He was observed reentering the lift station and, while climbing back up, collapsed and fell from the ladder into the accumulation of sewage in the bottom of the station. Employee #2, seeing Employee #1 collapsed, climbed down the ladder and got dizzy. While climbing back up the ladder he lost consciousness near the top. Employee #2 regained consciousness lying on top of the lift station. He observed Employee #3, an employee of the Public Service district, put on a body harness without a retrieval line and enter the lift station. Employee #3 collapsed and fell from the ladder. Employee #3 was retrieved from the bottom of the station by hooking the body harness with the hook of truck mounted hoist that was on the site. Employee #3 was transported to a hospital where he died the next day from hydrogen sulfide intoxication. Employee #2 was transported to a hospital, held for observation and released. Employee #1 was retrieved by a local emergency medical rescue squad. He was pronounced dead on the scene. The cause of death was hydrogen sulfide intoxication with drowning.

https://www.osha.gov/ords/imis/accidentsearch.accident_detail?id=200630960

September 2003 OSHA Database: 0 Fatalities, 1 Injury (Hospitalized)

On September 7, 2003, Employee #1, a truck driver, was opening a hatch cover at the top of a tanker truck to allow for make-up air to enter the tank during unloading operations. The tanker contained chicken blood, which had been loaded 17 hours earlier. Due to the nature of the cargo, the head space of the tank contained hydrogen sulfide gas, which is highly toxic. As Employee #1 opened the hatch the toxic gas was released, causing him to lose consciousness and fall approximately 10.5 feet to the ground. Employee #1 was hospitalized and treated for multiple cuts and lacerations. A detailed investigation revealed that the employer was at fault for failure to evaluate and control toxic exposures to employees and for failure to provide fall protection for employees working 10.5 feet above the ground, which they were cited.

https://www.osha.gov/ords/imis/accidentsearch.accident_detail?id=201504289

August 2003 OSHA Database: 1 Fatality, 1 Injury (Hospitalized)

Attempted Rescue

At approximately 2:45 p.m. on August 4, 2003, Employee #1 went in to a 36 in. diameter manhole, with Employee #2 watching, to change a sump pump at the bottom which had failed over the weekend. The manhole was 7 ft below ground elevation, and 1 ft above ground elevation. Employee #1 was overcome by methane and hydrogen sulfide gas. Employee #2 tried to pull Employee #1 out and was also overcome by the gases. Coworkers pulled Employee #2 out, but could not pull

Employee #1 out. The Hanover County Emergency personnel had to ventilate the manhole with a ventilation air blower before Employee #1's dead body could be removed. Employee #2 was hospitalized and released on August 6, 2003.

https://www.osha.gov/ords/imis/accidentsearch.accident_detail?id=200921195

June 2003 OSHA Database: 1 Fatality, 3 Injuries (Hospitalizations)

On June 27, 2003, Employees #1, #2, #3, and #4 of Clean Harbors Environmental Services, were working to fixate lead in incinerated ash through the addition of phosphorus pentasulfide. During the process, the four employees were exposed to high levels of what was presumed to be hydrogen sulfide gas. They had been wearing supplied air respirators, but were not wearing respiratory protection at the time of the incident. Employees #2, #3, and #4, who were on the second level of the ash building, were hospitalized. Employee #1, who was on the third level of the building, died.

https://www.osha.gov/ords/imis/accidentsearch.accident_detail?id=200052066

October 2002 OSHA Database: 2 Fatalities

Attempted Rescue

On October 10, 2002, Employees #1 and #2 and two coworkers, of the Electrical Technologies Corporation, were installing PVC pipes in a trench to run cable through it. This duct would then connect to the 10 ft deep by 3 ft 3 in. wide manhole located nearby. Water from the manhole was seeping into the trench through the window outlets that connected the two, and the crew decided to use a pump to purge the water from the manhole. The manhole opening was located 5 ft away, but the trench was dug up to the point of the manhole walls. Employee #2 entered the manhole to unclog the pump hose when he became overwhelmed by the atmospheric conditions and fell, face down, in the water. Employee #1, the foreman, went in to rescue him and also loss consciousness. When one of the coworkers tried to help them, he became dizzy and immediately exited the manhole. Employees #1 and #2 died of hydrogen sulfide exposure and were pronounced dead at the scene. The fire department took readings of the atmosphere and purged the manhole prior to sending in its rescue crew.

https://www.osha.gov/ords/imis/accidentsearch.accident_detail?id=200674083

August 2002 OSHA Database: 1 Fatality

Attempted Rescue

On August 27, 2002, Employee #1 and a coworker were installing a new gate and removing the old one from a stand box filled with manure water from a manure pond. The dairy owner ran the pump to draw water out of the box and reduce the water level below the opening port of the gate. When the water level had been lowered, Employee #1 entered the box, removed the old gate and cleaned the old sealing mortar from around the opening. He and the coworker then lowered the new gate into the box. Employee #1 reentered the box to position the gate, while the coworker and one of the dairy owners stood outside the box and held onto the gate from above. They were watching Employee #1 when he suddenly rolled onto his right side and landed, back down, in the water, with his head submerged. The coworker entered the box to rescue Employee #1, passed out momentarily, regained consciousness, and then climbed out of the box under his own power. Emergency responders were called, and Employee #1 was extricated from the stand box. He was transported to the hospital, where he was pronounced dead of the inhalation of hydrogen sulfide and possibly methane. Hydrogen sulfide levels were later measured at 218 PPM under conditions approximating those of the accident.

https://www.osha.gov/ords/imis/accidentsearch.accident_detail?id=201103017

August 2002 OSHA Database: 0 Fatalities, 1 Injury (Hospitalization)

At approximately 7:00 p.m. on August 25, 2002, Employee #1, a galvanizer at a wire manufacturing company, was pulling out wire from a gravel box that contained a mixer-burner through which the wire was drawn to produce a smooth finish. The burner was fueled by a mixture of natural gas and hydrogen sulfide, and Employee #1 was apparently struck in the face by a cloud of hydrogen sulfide. He was subsequently hospitalized for three days.

https://www.osha.gov/ords/imis/accidentsearch.accident_detail?id=201796455

July 2002 OSHA Database: 2 Fatalities, 0 Injuries

7/18/2002: Employees #1 and #2 were conducting a land survey to establish horizontal coordinates and elevations for a new sewer line. They entered a manhole leading to a 16 ft deep junction box, when they were overcome by sewer gas.

Some workers at a nearby manufacturing facility became aware of a rotten egg odor and began to search for the source. One worker peered through the open manhole and saw Employees #1 and #2. They died of asphyxiation from the hydrogen sulfide fumes.
https://www.osha.gov/ords/imis/accidentsearch.accident_detail?id=200372266

July 2002 OSHA Database: 2 Fatalities, 5 Injuries (All Hospitalized)

Attempted Rescue

On July 8, 2002, two employees worked in a 2-foot by 5-foot sanitary sewer conduit that was a permit-required confined space. The sewer conduit contained one to two feet of water. The employees were exposed to inhalation hazards such as hydrogen sulfide, unknown gases, and an oxygen-deficient atmosphere. They were overcome by the hazardous atmosphere and were asphyxiated. Five employees, who attempted to rescue their coworkers, were also overexposed to the hazardous atmosphere, and were hospitalized.
https://www.osha.gov/ords/imis/accidentsearch.accident_detail?id=200525608

January 2002 OSHA Database: 0 Fatalities, 8 Injuries (All Hospitalized)

On January 16, 2002, two employees were killed, and eight others were hospitalized due to an unexpected release of hydrogen sulfide. (The eight hospitalized employees were not listed on injury lines on the original form.) Note: Listed separately as water and sewer line construction.
https://www.osha.gov/ords/imis/accidentsearch.accident_detail?id=200451136

April 2001 OSHA Database: 0 Fatalities, 3 Injuries (Hospitalized)

Attempted Rescue

April 25, 2001: At approximately 2:00 p.m. on April 25, 2001, Employees #1, #2, #3 were exposed to hydrogen sulfide gas when they entered a confined space situation. All three employees worked at a municipal waste transfer station doing business as BFI Waste Services, which is owned by Allied Waste Industries, Incorporated. At first,

Employee #1, a facility supervisor, was seriously injured when he entered a flooded underground vault-drain to check out why it wasn't draining. While in the vault, Employee #1 was overcome with hydrogen sulfide gas, when he moved a loose fitting sewer pipe that crossed the underground confined space. The sewer pipe measured 8 inches. Employees #2 and #3, who worked under the facility supervisor's direction, were also seriously exposed to hydrogen sulfide when they attempted to rescue their supervisor. All three employees went to a local hospital for treatment.

https://www.osha.gov/ords/imis/accidentsearch.accident_detail?id=201171113

November 2000 OSHA Database: 1 Fatality, I injury

5900 Up River Rd

Corpus Christi, TX 78407

November 29, 2000: Employees #1 and #2, who worked for a nested maintenance contractor, were finishing the turnaround of the sulfur recovery complex at a refinery. They were removing a 14 in. isolation blind from the overhead inlet of a horizontal receiver vessel. The vessel was part of an amine treating unit that had been emptied, steamed out, and drained a few days before. After several attempts, the overhead piping had been replaced and the blinds had been removed and reinstalled. Employees #1 and #2 were working from a scaffold when they were exposed to strong hydrogen sulfide emissions from the vessel. Employee #1 staggered away, but within minutes had lapsed into unconsciousness and died. Employee #2 managed to escape and reach grade level. He was hospitalized for observation and released with no lasting effects. The vessel had accumulated sour gas from a connected overhead gas line, tied into nearby sulfur trains that were operating at relatively low pressure. The source was a single leaking 12 in. gate valve that had been closed and locked out. Employees #1 and #2 were working without respiratory protection or gas detection equipment. The valve inspection program, lockout/tagout program, and respiratory protections were found lacking. At the time of the accident, the foreman was also overseeing other crews at the site.

https://www.osha.gov/ords/imis/accidentsearch.accident_detail?id=200210755

October 2000 OSHA Database: 1 Fatality

Chemical Mixing

October 7, 2000: Employee #1 was working on the top of a reactor unit in a waste water treatment facility. He was checking the level of the solution in the reaction chambers through open port holes in the reactor. Waste acids in the reactor were to

be neutralized with waste caustics. A sulfide solution was accidentally added causing a release of hydrogen sulfide. Employee #1 was over exposed, collapsed, and died.
https://www.osha.gov/ords/imis/accidentsearch.accident_detail?id=200031144

July 2000 OSHA Database: 0 Fatalities, 1 Injury (Hospitalization)

At approximately 10 p.m. on July 31, 2000, Employee #1, the night supervisor, was observing the progress of a steam cleaning operation involving the out-of-service vapor recovery line to tank #1025. He was investigating the area around the 10 in. blind flange and the spreader bars, which was reportedly plugged, when a stream of water began to flow from the top of the flange. Employee #1 began to smell hydrogen sulfide gas and saw a stream of water and light hydrocarbons start to gush out of the area around the spreader bars and cover the 6 ft square scaffold platform. He later stated that he held his breath, climbed over the railing and began to descend the 21 ft scaffold ladder. Shortly thereafter, his hands and feet slipped from the rungs and he fell to the asphalt floor. Employee #1 suffered a fractured right femur, right knee, and left shoulder, and was hospitalized for 10 days at Long Beach Memorial Hospitalized. Among the causal factors that led to the accident were the lack of controls to safely dispose of the oil-water mixtures and the foreseeable toxic effects of hydrogen sulfide during maintenance activities. There were no bleed valves on the system to relieve pressure in the piping and no covers/diverters over the flanges to direct oily-water down instead of onto the ladder rungs.
https://www.osha.gov/ords/imis/accidentsearch.accident_detail?id=201033305

May 2000 OSHA Database: 1 Fatality

May 16, 2000: Between 8:00 a.m. and 8:30 a.m. on May 16, 2000, Employee #1, a process operator at the Cedar Rapids (IA) water pollution control facility, went atop of a tank full of sludge that contained hydrogen sulfide. He apparently opened the hatch to measure the amount of sludge in the tank and was exposed to levels of hydrogen sulfide greater than 300 ppm. At approximately 8:30 a.m., he was found dead by a coworker, who summoned help. Employee #1 was not wearing a respirator. There were no witnesses to the accident.
https://www.osha.gov/ords/imis/accidentsearch.accident_detail?id=200521755

February 2000 OSHA Database: 1 Fatality

February 27, 2000: Employee #1 was removing a sewage tank cover on the USS Peleliu at Pier 13 at the US Naval facility in San Diego, CA, when he was apparently overcome by hydrogen sulfide gas. He was killed by the gas. Employees were not required to wear respiratory protection while removing the sewage tank covers. There were no appropriate rescue procedures, no properly trained rescuers, and no easily available rescue equipment, such as harnesses, retrieval lines, and working supplied air respirators. All of these deficiencies contributed to Employee #1's death.

https://www.osha.gov/ords/imis/accidentsearch.accident_detail?id=201780202

February 2000 OSHA Database: 1 Fatality

Chemical Mixing

February 24, 2000: Employee #1 was adding sodium hydrosulfide to the magnesium chloride measuring tanks in order to improve the quality of the liquid polymer in the polymer reactor. Employee #1 had completed adding sodium hydrosulfide to the south tank and had started adding it to the north tank when he was overcome by hydrogen sulfide gas. Employee #1 died.

https://www.osha.gov/ords/imis/accidentsearch.accident_detail?id=201361136

January 2000, OSHA Database: 1 Fatality

January 1, 2000: On January 15, 2000, Employee #1 collecting brine samples from a 9-ft-tall, 1200-gallon fiberglass tank, fell approximately 6 ft off the ladder he was standing on while collecting a brine sample. Employee #1 hit his head on a concrete floor when he fell. He was hospitalized after the accident and died on January 27, 2000, from the head injuries. Employee #1 was working with brine containing hydrogen sulfide gas (H_2S). Concentrations of H_2S were measured after the accident and were 1000 ppm.

https://www.osha.gov/ords/imis/accidentsearch.accident_detail?id=302314505

November 1999 OSHA Database: 0 Fatalities, 3 Injuries (1 Hospitalized, 2 Released after treatment)

Attempted Rescue

At approximately 4:00 p.m. on November 11, 1999, three employees were injured while working in a confined space. Employee #1 was the stand-by attendant for a confined space entry for Employee #2 and Employee #3. The confined space was a shaft that was 18 ft deep and 5 ft in diameter. This was used as a manhole entrance to the juncture of a newly constructed 42 in. sewer line and an existing 30 in. live sewer line. As the two sewer lines were being joined, the inflatable "flow-through" plug being used to isolate the live sewer from the new sewer, was removed from the live sewer line. This resulted in sewage liquid and gases entering the shaft. The ladder used to get into the sewer was removed to facilitate the removal of the inflatable plug to the street surface using a backhoe to lift the plug. At that time the 32 in. fan being used to provide augmented ventilation to the confined space was turned off to take readings of the air contaminant concentration. The meter being used fell into the live sewer and was not recovered. After the plug was lifted to the street, the ladder was replaced into the shaft so that Employees #2 and #3 could exit. Employee #2 experienced "breakthrough" of the organic vapor cartridge air-purifying respirator being used and he removed the respirator. Employee #2 lost consciousness while ascending the ladder and was supported by Employee #3. Employee #1 then entered the shaft without respiratory protection or ventilating air, and descended the ladder approximately 3 ft to assist in extricating Employee #2. While on the ladder, Employee #1 lost consciousness and fell off the ladder to the bottom of the shaft. A fourth employee assisted Employee #3 in getting Employee #2 to the surface, where Employee #3 also lost consciousness. Employee #4 attempted to restart the fan while employees from another employer at the site conducted CPR on Employee #2. Once the fan was restarted, Employee #4 entered the shaft, without respiratory protection, using a rope to assist Employee #1 back to the surface. Employees #1, #2 and #3 were transported by ambulance to local hospitals. Employee #1 was admitted for a period four days and placed on mechanical ventilation life support. Employee #2 was released from the emergency room, against medial advisement, after receiving respiratory therapy. Employee #3 was released after approximately 4 hours after receiving respiratory therapy. Emergency Response Personnel collected readings for oxygen, LEL, carbon monoxide, and hydrogen sulfide in the shaft using a Biosystems PHD Ultra (Serial number 09142) multi- gas meter. Readings obtained within the shaft revealed 300 ppm hydrogen sulfide without mechanical ventilation. No elevated readings were obtained when the fan was running.

https://www.osha.gov/ords/imis/accidentsearch.accident_detail?id=201482692

November 1999 OSHA Database: 1 Fatality

On November 8, 1999, Employee #1 was found lying on a catwalk between two oil well storage tanks. It appeared that he may have died from exposure to hydrogen sulfide.

https://www.osha.gov/ords/imis/accidentsearch.accident_detail?id=200450906

November 1999 OSHA Database: 0 Fatalities, 1 Injury

At a approximately 10:55 a.m. on November 03, 1999, Employee #1 fell approximately 12 ft from a ladder while helping to install a section of a six-inch recovery line on a 2,000 barrel slop tank. Employee #1 was operating the crane utilized to support the section of the pipe. A coworker was acting as a safety watch and holding the ladder. Another coworker was on the ladder performing the maintenance with a breathing apparatus to protect him from potentially high levels of hydrogen sulfide. The coworker on the ladder came down to change air bottles when Employee #1 climbed the ladder to continue maintenance without proper protective equipment. Within minutes Employee #1 fell approximately 12 ft from the ladder. Employee #1 was hospitalized.

https://www.osha.gov/ords/imis/accidentsearch.accident_detail?id=170864664

September 1999 OSHA Database: 1 Fatality, 1 Injury (Hospitalization)

On September 4, 1999, there was an unplanned plant-wide power failure at the facility where Employees #1 and #2 worked. As a result, the #4 dryer at the plant's hydro unit started to decompose. Workers tried unsuccessfully to control the decomposition by flooding the dryer with water. Employees #1 and #2, both control room operators, entered the control room, apparently to reset some fans in anticipation of the power being turned back on. Assuming that the control room was a safe haven, they removed their SCBAs. When the power came back on, sulfur dioxide gases and some hydrogen sulfide entered the control room via the air conditioning and Purafil systems, whose duct works had holes. The two men were overcome by the fumes. Employee #1 died and Employee #2 was hospitalized. He could not speak and had lost control of his extremities, although he apparently understood what was going on and could nod yes and no.

https://www.osha.gov/ords/imis/accidentsearch.accident_detail?id=200450864

September 1999 OSHA Database: 1 Fatality, 1 Injury

September 4, 1999: Employee #1 and a coworker were overcome by hydrogen sulfide and collapsed. Employee #1 died of asphyxia.

https://www.osha.gov/ords/imis/accidentsearch.accident_detail?id=200450872

July 1999 OSHA Database: 1 Fatality, 1 Injury (Hospitalization)

Attempted Rescue

At 10:00 a.m. on July 26, 1999, Employee #1 was sent into a confined space to repair a malfunctioning valve. Employee #1 climbed a ladder to the top of the cement standpipe which stood 15 feet tall and is three feet in circumference. Employee #1 entered the confined space and was observed 20 seconds later trying to leave the confined space. He seemed to have a convulsion and lost his grip falling back into the pipe. A co-worker attempted to rescue Employee #1 but was unable to do so. A nearby independent contractor entered the confined space to rescue Employee #1 but was also overcome. The local fire department was able to extract both Employee #1 and the independent contractor. The independent contractor survived after being hospitalized for several days. Employee #1 died two days later. The coroner determined that Employee #1 and the independent contractor were overcome by hydrogen sulfide gas.

https://www.osha.gov/ords/imis/accidentsearch.accident_detail?id=201081767

July 1999 OSHA Database: 1 Fatality, 1 Injury

Mixing chemicals

Attempted Rescue

July 7, 1999: Employee #1 and a coworker were neutralizing phosphoric acid with ferrous sulfate and sodium sulfide inside a cement truck. The coworker was standing on top of the truck when he was overcome by hydrogen sulfide and slumped down on the truck's work platform. Employee #1 saw him and went to help, but he was also overcome by the chemical fumes. He was transported to the hospital, where he later died.

https://www.osha.gov/ords/imis/accidentsearch.accident_detail?id=200781466

June 1999 OSHA Database: 1 Fatality, 3 Injuries

Mixing chemicals

On June 4, 1999, Employee #1, a truck driver from Quality Carriers, Inc., was transferring sodium hydrosulfide into tanks of ferrous sulfate at Whitehall Leather Company. A chemical reaction occurred and hydrogen sulfide gas was generated and released into the workplace. Employee #1 was asphyxiated and died. At least three other employees from Whitehall Leather Company experienced varying degrees of symptoms from their exposure.

https://www.osha.gov/ords/imis/accidentsearch.accident_detail?id=201640026

December 1999 OSHA Database, 0 Fatalities, 10 Injured (4 Hospitalized)

Attempted Rescue

December 14, 1999: Employees #1 through #10 worked at a crude oil refinery (Citgo Refinery, Sru West Plant, Corpus Christi, TX 78469) at which nearby process units (crude oil and gas oil treater) were undergoing a planned turnaround. The units had staggered shutdown and startup schedules, and they shared some product transfer piping leading to storage. When the product flow was rerouted by operations prematurely, sour naphtha liquid was released into the crude unit from an open connection installed on a new 6 in. piping section that served as an exchanger bypass line. The low-pressure release lasted for several minutes and allowed clouds of toxic hydrogen sulfide gas to accumulate in and around the nearby control room and through nearby process units. Emergency alarm and evacuation procedures were not executed according to plan. Operators responding in the field at first attempted rescue efforts without using emergency air packs, ultimately adding to the total of exposed employees. Numerous exposures to the evolved sour gas resulted in injury to both plant and contractor employees. Employees #1 through #10 were transported to the hospital, and Employees #2 through #4, #9, and #10 were admitted. None experienced permanent effects. Subsequent investigation revealed weaknesses and discrepancies in the plant lockout/tagout program, and poor implementation and upkeep of the process safety program. Unit isolation blinds and lockout/tagout blinds for the equipment were not clear to the operators due to poor tracking. Operating procedures were both outdated (e.g., the use of checklists) and/or unclear, such as to how to perform the final stages of unit startup.

https://www.osha.gov/ords/imis/accidentsearch.accident_detail?id=200210391

December 1998 OSHA Database: 0 Fatalities, 10 Injured (Treated at Hospital and released)

December 9, 1998: Employees #1 through #10 were at work when a power failure occurred at the company. The company's uninterruptible power system (UPS) failed and the battery system overheated. The smell of hydrogen sulfide was noticeable in parts of the building. The employer tried unsuccessfully to restart the UPS, and then called 911. The fire department responded and gave evacuation orders at 10:40 a.m. Employees #1 through #10 were transported to the hospital, where they were treated to exposure to hydrogen sulfide and battery acid, and released.

https://www.osha.gov/ords/imis/accidentsearch.accident_detail?id=200600518

October 1998 OSHA Database: 1 Fatality, 2 Hospitalizations, 1 Treated and released

October 2, 1998: Employees #1 through #4 were in a fishing void when they were overcome by hydrogen sulfide and a lack of oxygen. Employee #1 fell into waste water at the bottom of the confined space and drowned; the causes of death were listed as both hydrogen sulfide poisoning and drowning. Employees #3 and #4 were hospitalized; Employee #2 was treated and released.

https://www.osha.gov/ords/imis/accidentsearch.accident_detail?id=200460350

September 1998 OSHA Database: 1 Fatality, 4 Injuries (All Hospitalized)

September 21, 1998: Some employees were installing a French drain system to collect water seeping from a slurry pond. The employees were entering the catch basin to do the final touch-up work by riding the bucket of a backhoe down into the basin. One of the employees, a 57-year old supervisor, was engulfed by vapors that were later found to be hydrogen sulfide. He died of inhalation of toxic fumes. Four other employees were hospitalized for exposure to the hydrogen sulfide.

https://www.osha.gov/ords/imis/accidentsearch.accident_detail?id=200370245

September 1998 OSHA Database: 1 Fatality, 6 Injuries (Including 3 Hospitalizations)

Attempted Rescue

On September 17, 1998, several employees were retrieving and repairing a chain in a pit that contained water and inedible catfish parts. Employee #1 was exiting the pit when he fell back into the pit. Employee #2 was holding Employee #1's head to keep it above the material in the pit when he collapsed. Employee #3 and Employee #4 attempted to enter the pit, but were unable to do so. Both Employees #3 and #4 collapsed outside of the pit. Employee #5 attempted to enter the pit, but was unable to do so, and collapsed outside the pit. Employee #6 entered the pit and collapsed inside the pit. Employee #6 was rescued by Employee #7. Employee #1 was asphyxiated by hydrogen sulfide gas and died. Employees #2, #6 and #7 were admitted to the hospital. Employees #3, #4 and #5 were not hospitalized. The inspection indicated that permit-required confined space procedures were not followed; an entry permit was not completed, atmospheric testing not performed, retrieval equipment was not available, and personal protective equipment was not utilized.

https://www.osha.gov/ords/imis/accidentsearch.accident_detail?id=201360625

September 1998 OSHA Database: 1 Fatality

September 1, 1998: Employee #1 entered a 9 ft deep manhole to apply sealant to the connecting concrete rings. This was the last, and the deepest, of the six manholes he had entered. Shortly after reaching the bottom, Employee #1 was overcome by hydrogen sulfide gas that had collected in the manhole. He was killed.

https://www.osha.gov/ords/imis/accidentsearch.accident_detail?id=200210631

August 1998 OSHA Database: 2 Fatalities

Attempted Rescue

August 19, 1998: Employees #1 and #2 were part of a construction crew building an extension sewer line that was to tap into an existing city line. The crew had exposed one side of a manhole in the city sewer line and a subcontractor had core-drilled a hole in it for placement of the new line. Some concrete remained intact after the drilling was completed. Employee #1 was lowered into the manhole using a chain draped over a rock bar. He was immediately overcome by the high levels of hydrogen sulfide. Employee #2 attempted to rescue him but was also overcome by the fumes. Both workers were killed.

https://www.osha.gov/ords/imis/accidentsearch.accident_detail?id=200710630

July 1998 OSHA Database: 2 Fatalities

Attempted Rescue

July 28, 1998: Employee #1, assigned to clean out a tanker trailer which had contained waste water from a pet food manufacturing facility, entered the tanker and collapsed. Employee #2 who was watching Employee #1 went inside to get help. Employee #2 and two other employees returned to the tanker trailer and Employee #2 entered the tanker trailer and collapsed. The other two employees went inside the facility and summoned emergency medical services and the Plant Superintendent. Employee #1 and Employee #2 were asphyxiated by hydrogen sulfide gas and died. Permit-required confined space procedures had not been followed and an entry permit was not completed. The air was not tested for oxygen deficiency or toxic atmosphere nor was retrieval equipment utilized. The only personal protective equipment worn by the employees were gloves, boots, and a hard hat.

https://www.osha.gov/ords/imis/accidentsearch.accident_detail?id=201340320

February 1998 OSHA Database: 1 Fatality

On February 11, 1998, Employee #1 was on top of a sour water tank blocking in steam eductor valves when he received an acute exposure to hydrogen sulfide. He was on top of the tank alone for about 10 minutes without any additional PPE. He was wearing only flame retardant clothing and a hard hat. Safety procedures require the use of a self-contained breathing air respirator and the necessary breathing air equipment was present on the tank. Hydrogen sulfide guidelines require that an employee must have a respirator on when at the top of the stairway of the tank. Employee #1 was still alive on February 17, 1998, the date this incident was reported to the OSHA office, but his brain activity was only 8 percent of normal. He was on complete life support, and it was very questionable whether he would live. Employee #1 died on February 22, 1998, at John Sealy Hospital - Texas City Medical Branch.

https://www.osha.gov/ords/imis/accidentsearch.accident_detail?id=201921285

February 1998 OSHA Database: 1 Fatality

February 8, 1998: At approximately 10:00 p.m. on February 8, 1998, Employee #1 was preparing to launch a pipeline scraper that was to travel the propane pipeline from the Star Enterprise Louisiana plant to the Sorrento, LA, storage facility. He had apparently removed the vent value plug, opened the vent valve to the pressure gauge, and closed the pressure gauge isolation valve. He had removed the pressure warning lock from the metal tab that extended from the outside edge of the launcher door over the launcher housing; unscrewed the launcher door, moved it aside, and inserted the pipeline scraper, and positioned it for launching; and greased the threads for the launcher door. Employee #1 started screwing the door back on the launcher when he was overcome with hydrogen sulfide that had backed up from the flare system. He was found unconscious on the ground and died of asphyxia. Apparently Employee #1 had not closed the depressurizing valve to the flare before opening the launcher door. The launcher had been depressurized to zero pressure and the flare line had a positive pressure of 3 psi. Employee #1 was not using any form of respiratory protection.

https://www.osha.gov/ords/imis/accidentsearch.accident_detail?id=200550135

July 1997 OSHA Database: 3 Fatalities

July 12, 1997: Employees #1 through #3, all shipyard workers, were overcome by hydrogen sulfide and/or methane gases from raw liquid sewage that leaked into the collection, holding, and transfer (CHT) pump room in which they were working. The three employees were rendered unconscious and subsequently drowned in the sewage.

https://www.osha.gov/ords/imis/accidentsearch.accident_detail?id=200220044

April 1997 OSHA Database: 0 Fatalities, 1 Injury (Treated at hospital and released)

At approximately 2:00 p.m. on April 10, 1997, Employee #1, a vacuum truck driver for Hayter Trucking, Inc., was pulling oil out of a drain pit north of the AFS office at Mobil Oil in Fellows, CA. He was working with a crew from Ken Small construction, and using three monitors designed to detect hydrogen sulfide. The monitors' alarms sounded and the foreman for Ken Small told everyone to leave the area. Employee #1 pulled the truck hose out of the pit and went to the back of his truck to close a valve. He told the foreman he had inhaled a small amount of hydrogen sulfide and wasn't feeling well. Employee #1 was taken to a safe area, given a drink of water, and then transported to Westside District Hospital, where he was treated and released.

https://www.osha.gov/ords/imis/accidentsearch.accident_detail?id=170894901

October 1996 OSHA Database: 0 Fatalities, 13 Injuries including 3 Hospitalizations

Mixing chemicals

On October 17, 1996, several employees were cleaning and refilling drums. They used a smell test to determine its contents. This led to the mixing of incompatible substances in the waste pit, which resulted in a toxic gas such as hydrogen

sulfide release. This incompatible mixture caused eight employees and later five firefighters to be taken to the hospital. Five of the employees and the five firefighters were treated and released the same day. The other three employees were hospitalized overnight. Employees stated there was a rotten egg odor and DEP reported that Hydrogen Sulfide was detected at the site at 35 ppm on October 17.
https://www.osha.gov/ords/imis/accidentsearch.accident_detail?id=200831030

September 1996 OSHA Database: 2 Fatalities

September 28, 1996: Employees #1 and #2 were draining chemicals from plating tanks when they inhaled lethal amounts of hydrogen sulfide. Both were killed
https://www.osha.gov/ords/imis/accidentsearch.accident_detail?id=200690063

September 1996 OSHA Database: 4 Fatalities

Attempted Rescue

September 5, 1996: Employee #1 entered a confined space to remove a bypass hose. When he detached the hose, sewage and waste water entered the space, and the toxic atmosphere caused him to lose consciousness. Employees #2 through #4 immediately entered the confined space to rescue Employee #1. They were overcome with hydrogen sulfide gas, lost consciousness, fell into the sewage water, and drowned. All four employees were killed.
https://www.osha.gov/ords/imis/accidentsearch.accident_detail?id=14509442

August 1996 OSHA Database: 2 Fatalities

Attempted Rescue

August 9, 1996: Employees #1 and #2, both riggers, were capping a sulfur well in a shallow bay. They were working from the deck of a barge equipped with a crane with a clamshell bucket. The employees dug around the 10-inch diameter well casing, then set a 10-foot diameter by 20-foot long caisson around the wellhead. Standard procedures required them to cut off the 10 inch casing approximately 15 feet below the line, then weld a circular plate over the end. Employee #1 went into the caisson to wrap a sling around the pipe after they had cut it off with a pneumatic saw. He was overcome by hydrogen sulfide gas. Employee #2 entered the caisson to rescue Employee #1, and was overcome also. Both were killed. Neither man was wearing a respirator.
https://www.osha.gov/ords/imis/accidentsearch.accident_detail?id=170081962

March 1996 OSHA Database: 0 Fatalities, 2 Injuries (1 Hospitalized)

Mixing Chemicals

March 4, 1996: Employees #1 and #2 were using muriatic acid to flush out an obstruction in some plumbing that carried green liquor. They added 300 gallons of acid to the pipes, and as the acid combined rapidly with the liquor, it triggered a violent chemical reaction that involved heat, chemical vapors of hydrogen sulfide, and boiling fluids. Both Employees were overcome by the vapors. Employee #1 suffered immediate asphyxia that affected his pulse and respiration, and lost consciousness. Employee #2's exposure was delayed and less severe; his injuries were the result of falling to the ground.
https://www.osha.gov/ords/imis/accidentsearch.accident_detail?id=171038037

February 1996 OSHA Database: 2 Fatalities

Attempted Rescue

February 28, 1996: Employees #1 and #2 were removing a temporary pump and water line from a water flood system at an oil well water injection plant. They were preparing to re-install the permanent pump, which had been taken out for repairs. The employees had not bled the pipe system of water because they were only going to remove PVC line from the dead end of the line immediately downstream of a closed valve. Employee #1 apparently removed a vise clamp immediately upstream of the closed valve, causing the water line to separate and water from the pipes and from a 40-barrel outside holding tank to enter the building. Hydrogen sulfide was released from the water, and Employee #1 was overcome by the fumes. Employee #2 was apparently overcome when he attempted to rescue his coworker. Both employees were killed.
https://www.osha.gov/ords/imis/accidentsearch.accident_detail?id=14419568

January 1996 OSHA Database: 0 Fatalities, 3 Injuries (All Hospitalized)

At approximately 8:15 a.m. on January 27, 1996, Employee #1 was bleeding an accumulator tank, a task he routinely performs several times each day. These tanks accumulate condensate of sour water, rich in hydrogen sulfide. He had grabbed a 3/4 in. ball valve to bleed the tank when the bleed pipe broke. The sour water blew out under 260 to 270 psi, hitting a concrete wall approximately 16 in. away, engulfing the area in sour water, and reducing visibility around the immediate area. In attempting to shut off an up-line valve, Employees #1 through #3 were exposed to hydrogen sulfide gas and were hospitalized.
https://www.osha.gov/ords/imis/accidentsearch.accident_detail?id=171016116

January 1996 OSHA Database: 2 Fatalities

Attempted Rescue

On January 23, 1996, Employee #1 entered an asphalt tanker after observing water at the bottom of the tanker. He entered without a respirator to remove the water with a bucket and sponge. Shortly afterwards, he felt weak, shouted for help, and went motionless. After being informed of the situation, Employee #2 entered the tanker in order to rescue Employee #1. He entered the tanker without a respirator. Employee #1 and Employee #2 were exposed to very high concentrations hydrogen sulfide inside the asphalt tanker and were killed.
https://www.osha.gov/ords/imis/accidentsearch.accident_detail?id=201310331

January 1996 OSHA Database: 1 Fatality

Well shutdown and audible alarm bypassed

Employee #1, owner of Graves Production Service, was contracted by Bishop Petroleum, Inc., to gauge its tanks daily. At approximately 8:30 a.m. on January 14, 1996, Employee #1 went to gauge the tanks. At approximately 9:00 p.m., neighbors complained of a strong smell coming from Bishop Petroleum. The volunteer fire department responded and at approximately 9:30 p.m. found Employee #1 dead, lying on his back on top of a tank. Both vents on the tank were open and both hydrogen sulfide alarms (red lights) were on. Employee #1 was not wearing the required supplied air respirator. He had bypassed the well shutdown mode and the audible alarm. The alarm was activated when a level of 20 ppm of hydrogen sulfide was detected. The shutdown mode was activated when a level of 50 ppm was detected.
https://www.osha.gov/ords/imis/accidentsearch.accident_detail?id=170582167

January 1996 OSHA Database: 3 Fatalities, 2 Injuries (Hospitalizations)

Attempted Rescue

January 4, 1996: Employee #1 was inside a frac tank shoveling residue (called BS) to one end of the tank for subsequent vacuuming and removal. As he completed the task, the tank was washed down with waste water containing hydrogen sulfide. Approximately 8 minutes following the waste water entrance, Employee #1 collapsed from exposure to the chemical. Employee #2 entered the tank and attempted to rescue Employee #1, but he too collapsed. Apparently Employees #3 and #4 entered the tank and attempted a rescue and but succumbed also. Employee #5 attempted to revive Employee #1 through an opening at the end of the frac tank. He was affected by the hydrogen sulfide gas but was able to call the city's first responders. Employees #1, #3, and #4 died of hydrogen sulfide exposure. Employees #2 and #5 were hospitalized.

https://www.osha.gov/ords/imis/accidentsearch.accident_detail?id=559229

December 1995 OSHA Database: 1 Fatality

December 13, 1995: Employee #1 was lowered into a trench containing crude oil and hydrogen sulfide gas. He died of asphyxiation due to the high concentration of hydrogen sulfide gas.

https://www.osha.gov/ords/imis/accidentsearch.accident_detail?id=880948

October 1995 OSHA Database: 1 Fatality

October 10, 1995: Employee #1 was repairing the rod of a gauge inside a Bobtail tank truck. Fifteen percent oxygen was present. Toxic substances present were hydrogen sulfide, 6 percent, carbon monoxide, 9 percent, and residues of propane. Employee #1 entered the tank without the required entrance permit for confined spaces. He died of oxygen deficiency. Employee #1 had not been trained in the procedures, hazards, and permit requirements to enter confined spaces. The employer did not post signs warning about unauthorized entrance into this confined space by either employees or others.

https://www.osha.gov/ords/imis/accidentsearch.accident_detail?id=170075899

September 1995 OSHA Database: 1 Fatality

September 15, 1995: Employee #1 was taking samples of molten sulfur from a tank truck. No respirator was provided. He was either standing or kneeling on a platform attached to the tanker truck 8 ft 10 in. from the ground. He had opened the manway of the tank and leaned across the manway, reaching approximately 15 in. into the tank to obtain samples of molten sulfur. At this point, it appears that he was overcome with hydrogen sulfide gas and fell to the gravel-covered ground. He was found unconscious within 10 minutes by coworkers and was pronounced dead shortly thereafter. The autopsy listed the cause of death as blunt force injuries. OSHA personal sampling of the same conditions showed hydrogen sulfide readings of up to 190 ppm. Company sampling under conditions where employees wore SCBA and stood on the ladder, second rung from the top, showed readings up to 809 ppm. OSHA maintains that hydrogen sulfide was a causal factor of the fall.

https://www.osha.gov/ords/imis/accidentsearch.accident_detail?id=170087944

September 1995 OSHA Database: 0 Fatalities, 3 Injuries (All Hospitalized)

September 14, 1995: Employee #1 inhaled fumes of leaking hydrogen sulfide gas and began to feel dizzy. Employees #2 and #3 tried to rescue him, but they were also overcome by the fumes. All three employees were hospitalized.

https://www.osha.gov/ords/imis/accidentsearch.accident_detail?id=724922

February 1995 OSHA Database: 1 Fatality

Work in Proximity

February 26, 1995: Employee #1, of Thomson Mechanical Corp., was connecting pipelines to condensate system pumps during a refinery turnaround at FCCU 152. At the same time, another contractor's crew was removing a defective sight level gauge from a propane relief drum in a scrubber system. The product inside the gauge spilled onto the ground. Employee #1 complained of a tight chest and was hospitalized. He died later at the hospital. The autopsy report listed hydrogen sulfide poisoning as the immediate cause of death.

https://www.osha.gov/ords/imis/accidentsearch.accident_detail?id=171005887

February 1995 OSHA Database: 3 Injuries (None Hospitalized)

February 2, 1995: A small gas line failed, resulting in a natural gas release outside a factory. Fresh air intakes near the ground pulled the gas, which included mercaptan and a small amount of hydrogen sulfide, into the building. Employees were told to go home at 12:00 noon. Employees #1, #2, and #3 experienced nausea, headaches, dizziness, and vomiting. None were hospitalized. This factory has experienced three natural gas problems in five years.

https://www.osha.gov/ords/imis/accidentsearch.accident_detail?id=170672364

November 1994 OSHA Database: 1 Fatality

Mixing chemicals

November 3, 1994: A contractor was hired to come on site and "acidize" a clogged water disposal well. The well was located in a 7 ft deep concrete vault. To unclog the well, the contractor put a hydrochloric acid mixture in the well's tubing. He warned the company that this process had the potential to form hydrogen sulfide in the well. After the contractor left the site, Employee #1 entered the well to replace a broken valve. When he broke the seal on the valve, hydrogen sulfide was released into the confined space and Employee #1 was asphyxiated. Coworkers found him later slouched over in the well vault. The autopsy concluded that he died of hydrogen sulfide intoxication. The well was not a permit-required confined space.
https://www.osha.gov/ords/imis/accidentsearch.accident_detail?id=552265

October 1994 OSHA Database: 1 Fatality

October 1, 1994: Employee #1 was checking fluid levels in tanks. When he opened a valve, hydrogen sulfide was released. Employee #1 inhaled the gas, was asphyxiated, and died.
https://www.osha.gov/ords/imis/accidentsearch.accident_detail?id=755314

July 1994 OSHA Database: 0 Fatalities, 2 Injuries (Both Hospitalized)

July 21, 1994: Employees #1 and #2 were exposed to and asphyxiated by hydrogen sulfide gas, which leaked past the packing of a valve that was being closed. While Employee #2 was rescuing Employee #1, he was overcome by the hydrogen sulfide and fell, fracturing his skull. Both employees were hospitalized.
https://www.osha.gov/ords/imis/accidentsearch.accident_detail?id=880567

June 1994 OSHA Database: 1 Fatality

Mixing chemicals

A black substance accumulates in the pipelines of the injection wells that are used in an oil field water-flood operation and occasionally has to be removed. Hydrochloric acid (HCl) is pumped through the pipelines to dissolve the black matter. On June 30, 1994, Employee #1, the employer, pumped HCl through a pipeline that was 3,500 to 4,000 feet long. The acid was taken to the location in one tank truck and another tank truck was positioned to receive the acid. As the acid passed through the pipeline, it reacted with the calcium sulfide and/or iron June 30, 1994: sulfide in the black matter and created hydrogen sulfide (H_2S). After the acid had been pumped through the pipeline and into the tank truck, the pump on the receiving tank truck was used to suck any remaining fluid from the hose. The pump evacuated H_2S from the tank and out of the tank vent. The H_2S created a lethal concentration outside the vent. Employee #1, the truck driver, walked into the vapor and immediately lost consciousness and died. Employee #2 attempted to assist his boss and was overcome by the toxic gas but recovered.

https://www.osha.gov/ords/imis/accidentsearch.accident_detail?id=772384

June 1994 OSHA Database: 1 Fatality, 2 Injuries (1 in Long term care, 1 treated and released)

June 14, 1994: Employees #1, #2, and #3 were in a dry well modifying sewer mains. Fluids left in the pipe for three months flowed into the work area. The fermenting fluids released hydrogen sulfide gas. Employees #1 and #2 were hospitalized. Employee #1 died of asphyxiation. Employee #2 is in a long term health care facility in Westchester, NY. Employee #3 was treated and released.

https://www.osha.gov/ords/imis/accidentsearch.accident_detail?id=170401178

May 1994 OSHA Database: 1 Fatality

Employee #1 had entered a tanker to perform work such as, but not limited to, cleaning it out, when he was overcome by hydrogen sulfide. The tanker had been used to haul rotten eggs. Employee #1 was killed by the fumes.

https://www.osha.gov/ords/imis/accidentsearch.accident_detail?id=170789184

April 1994 OSHA Database: 1 Fatality

April 6, 1994: Employee #1 was draining a fuel gas knockout drum in the hydrotreating unit of a petroleum refinery. He was overcome and killed by hydrogen sulfide gas.

https://www.osha.gov/ords/imis/accidentsearch.accident_detail?id=882605

March 1994 OSHA Database: 0 Fatalities, 4 Injuries (All four treated and released)

March 31, 1994: At approximately 8:30 a.m. on March 31, 1994, a strong noncondensible gas main line to the #12 boiler experienced a pressure buildup when the main gas line water loop seal drained into an open sewer. Employee #3 was in the vicinity of the drain and observed that his hydrogen sulfide monitor's alarm had been triggered. He asked Employee #2 to come over to verify his hydrogen sulfide monitor's reading. Employee #2 went to the control room to get pencil and paper to log the readings and on his return to the leak location told Employee #1, his supervisor, that the monitor's alarm had been triggered. Employee #1 left the control room to follow up by checking on water loop seals on the deck above. Noticing an odor and observing Employees #2 and #3 by the drain, he told them to leave. He attempted to fill the water loop seal twice and went into the gas leak area twice to see if he could see water flowing into the sewer from the loop seal drains. He left the area and called the kiln operator to shut off the gas to the boiler but could not reach him by radio. Employees #1, #2, and #3 reported to the medical center suffering inhalation symptoms. Employee #4 went to a doctor at about the same time with similar inhalation symptoms. Two employees were sent to the hospital and treated with oxygen before being released.

https://www.osha.gov/ords/imis/accidentsearch.accident_detail?id=170041925

February 1994 OSHA Database: 1 Fatality

February 10, 1994: Employee #1 was attempting to pump production water down the casing side of a well. He shut off the valve to gas gathering and removed a connection valve in order to place a 3 by 2 pipe swage and hook up the 3 in. line. When the connection valve was removed, hydrogen sulfide was released. Employee #1 died of asphyxia.

https://www.osha.gov/ords/imis/accidentsearch.accident_detail?id=14413710

February 1994 OSHA Database: 1 Fatality

Shell Oil Company Hwy 225(A) Center St. Deer Park, TX

February 3, 1994: At approximately 10:00 p.m. on February 3, 1994, Employees #1, #2, and #3 were conducting maintenance activities in the Gerbitol unit at DPMC. The employees were installing a #005 blind in the flange on the pipeline from

V-1110 (amine reflux accumulator). As the flange was opened by Employee #2, the hydrogen sulfide unit alarm sounded. The employees stopped their work. Employee #1 tried to remove his full face piece and was overcome by hydrogen sulfide gas, lost consciousness, and died. Employee #2 had previously loosened his strap to test the atmosphere for hydrogen sulfide in their work area. Employee #3, on ground level, went up to the work area where his 30 minute Scott air pack hung up, exposing him to the gas. Employees #2 and #3 were hospitalized.

https://www.osha.gov/ords/imis/accidentsearch.accident_detail?id=14221055

January 1994 OSHA Database: 2 Fatalities, 4 Injuries (All Hospitalized)

Attempted Rescue

January 26, 1994: Employees #1 and #2 entered a manhole to attempt repairs and died. Employee #3 attempted to rescue Employee #2 and was overcome by the hydrogen sulfide fumes. Employee #4 was overcome by fumes while attempting to rescue Employee #3. Employee #5 was overcome by fumes while attempting to rescue Employee #4. Employees #4 and #5 exited the manhole. The local fire department rescued Employee #3. Employee #6 became ill while outside the manhole.

https://www.osha.gov/ords/imis/accidentsearch.accident_detail?id=14360911

December 1993 OSHA Database: 1 Fatality

December 27, 1993: Employee #1, an oil field worker, fell approximately 20 ft when he was overcome by hydrogen sulfide gas while attempting to gauge a portable tank. He struck his head on frozen ground and died of a massive head injury.

https://www.osha.gov/ords/imis/accidentsearch.accident_detail?id=880682

November 1993 OSHA Database: 1 Fatality, 5 Injuries (All Hospitalized)

Attempted Rescue

On November 29, 1993, Employees #1 and #2, mechanics at Kopper's Industries, Stickney, IL, had ascended a coal tar pitch distillation unit to open a manway. The mechanics were to remove all but four bolts on each of three manways. The last four bolts were to be removed by another crew at which time air-supplied respirators were to be used. Apparently, Employees #1 and #2 decided to remove more than the usual number of bolts, causing the manway gasket seal to break open and emit hydrogen sulfide gas. The mechanics were not wearing respirators at the time nor were respirators in their possession. Both mechanics were overcome by the hydrogen sulfide gas, asphyxiating Employee #1 and rendering Employee #2 unconscious. In the rescue attempt, Employees #3 through #6, the rescuers, became ill from the gas due to lack of proper respirators and possibly from improper use of respirators. Employees #1 through #6 were taken by paramedics to the hospital: Employee #1 was dead on arrival. The remaining employees were hospitalized overnight; one employee remained for several days. The investigation revealed improper procedure by Employees #1 and #2.

https://www.osha.gov/ords/imis/accidentsearch.accident_detail?id=724237

November 1993 OSHA Database: 2 Fatalities, 8 Injuries (5 Were Hospitalized)

At approximately 9:30 a.m. on November 11, 1993, Employee #1 entered a Batter cylindrical tank (49 in. diameter, 473 in. length) through a 14 in. diameter hatch at the top. The tank, which held residue of molten sulfur, was brought into the shop for repair of the valves. Employee #1 probably entered the tank to examine the valves from within, as is the usual practice. Employee #5 was looking for the forklift when he discovered Employee #1 in the tank and went for help. Employee #2 entered the tank, rescued Employee #1, and was overcome by hydrogen sulfide. Employees #3 and #5 performed CPR to Employee #1, while Employee #4 tried to rescue Employee #2. Employee #8 pulled Employee #4 from the tank by his belt. Employee #4 passed out. Employees #7 and #8 tried to rescue Employee #2 from the tank. Employees #8 and #9 used a hook to pull Employee #2 out by his shirt sleeve. Employees #10 and #9 helped pull Employee #2 out. All 10 employees suffered from overexposure to hydrogen sulfide gas. Employee #1 died on November 11, 1993; Employee #2 died on November 15, 1993.

https://www.osha.gov/ords/imis/accidentsearch.accident_detail?id=14244842

June 1993 OSHA Database: 0 Fatalities, 2 Injuries (Both Hospitalized)

At approximately 8:15 a.m. on June 3, 1993, a floor drain in the radiology lab became plugged. An unauthorized employee put drain cleaner in the drain, which resulted in the release of irritating vapors throughout the lab. The facility was evacuated and the problem was resolved. Based on descriptions of the odors by the maintenance employee and a written report of the incident by the lab, one of the vapors may have been hydrogen sulfide. Employees #1 and #2 were hospitalized for four days for observation and treatment. A regulatory general citation was issued for not reporting a serious injury within 24 hours (342A). A general violation was issued relative to MSDS-5194(g)(8) and a serious violation was issued relative to Training-94(h)(1).

https://www.osha.gov/ords/imis/accidentsearch.accident_detail?id=170355036

March 1993 OSHA Database: 1 Fatality

March 13, 1993: Employee #1 manually opened a bleeder valve during normal operating conditions. The employee was found lying unconscious on the platform near the bleeder valve. He was exposed to the 153 lb of hydrogen sulfide, 1,008 lb of hydrocarbon, 5.8 lb of water vapor, and 37.6 gallons of sour water that was released. Employee #1 died.

https://www.osha.gov/ords/imis/accidentsearch.accident_detail?id=14410658

March 1993 OSHA Database: 1 Fatality, 2 Injuries (1 Was Hospitalized)

March 1, 1993: Employee #1 was killed and Employee #2 was injured when hydrogen sulfide gases were released into the air after they and Employee #3 removed dome lids that covered railroad cars containing molten sulfur. Employees #1 and #2 were working from an elevated motorized platform lift. Employee #3 was working at ground level attaching steam lines to the bottom of the rail car. Employee #1 opened the domes. Employee #3 found Employees #1 and #2 unconscious on the platform floor. He drove the platform away from the rail line and summoned help. Employee #1 was dead on arrival at a local hospital and Employee #2 was hospitalized for several days. Employee #3 was examined, then released.

https://www.osha.gov/ords/imis/accidentsearch.accident_detail?id=14339071

February 1993 OSHA Database: 0 Fatalities, 2 Injuries

February 8, 1993: three electric utility employees were involved in a geothermal drilling operation. They were preparing a well clean out of KS-9, which included a 4-hour venting. The steam used in this operation is normally abated with the use of caustic soda in the atmospheric separator and muffler system. However, during this particular operation, condensate followed by a cloud of unabated hydrogen sulfide came out of the sample port. Two employees were overcome by the gas. One of them had been standing near the throttle valve, and the other had been monitoring hydrogen sulfide from a nearby location.

https://www.osha.gov/ords/imis/accidentsearch.accident_detail?id=744409

February 1993 OSHA Database: 0 Fatalities, 2 Injuries

February 8, 1993: a consultant for a geothermal electric utility was standing at a wellhead near a 2-inch bypass. The gas cap came to the surface and flooded the area with a substantial amount of hydrogen sulfide. The consultant ran and collapsed about 150 feet northwest of the wellhead. He did not lose consciousness, however, and quickly regained his balance. A chemical technician, who was performing daily air monitoring, was also overcome by the hydrogen sulfide release. He donned a self-contained breathing apparatus near the ground, but collapsed before he could turn the air on. He was revived with oxygen. Both employees recovered from their injuries.

https://www.osha.gov/ords/imis/accidentsearch.accident_detail?id=744318

January 1993 OSHA Database: 1 Fatality

H_2S Explosion/Fire

Mixing Chemicals

On or about January 28, 1993, Employee #1, a contract welder, was repairing and replacing flanges on storage tanks in order to install a closed ventilation system. The storage tanks contained sodium sulfide, which reacts with acid to generate hydrogen sulfide gas. This highly flammable gas accumulated in a vapor space. Prior to the accident, a tank flashed while being cut or welded into by the employee. When a lighted torch was brought in proximity, the tank ruptured. Employee #1 died. Employee #2 had gone up on a tank to tell Employee #1 not to cut into the tank when the tank exploded. Employee #2 was hospitalized.

https://www.osha.gov/ords/imis/accidentsearch.accident_detail?id=14219463

January 1993 OSHA Database: 0 Fatalities, 1 Injury

On January 22, 1993, Employee #1 and several coworkers were pulling a geothermal drill pipe from its hole when they noticed that the pipe was plugged. Normally, mud is circulated through this system, but in this case, the one-way valves were later discovered to have failed. While the last few sections of the pipe were being removed, the surrounding air was checked for hydrogen sulfide using detector tubes. At this time, according to Employee #1, the hole started spitting out black sulfur water and hydrogen sulfide. When Employee #1 bent over to set the slips, he got a whiff of the hydrogen sulfide, felt light-headed, and was overcome. He was given oxygen and felt better. He was not hospitalized.

https://www.osha.gov/ords/imis/accidentsearch.accident_detail?id=744391

Link to the OSHA Database

https://www.osha.gov/ords/imis/AccidentSearch.search?acc_keyword=%22Hydrogen%20Sulfide%22&keyword_list=on

https://www.osha.gov/ords/imis/accidentsearch.search?sic=&sicgroup=&naics=&acc_description=&acc_abstract=&acc_keyword=%22Hydrogen%20Sulfide%22&inspnr=&fatal=&officetype=&office=&startmonth=&startday=&startyear=&endmonth=&endday=&endyear=&keyword_list=on&p_start=&p_finish=60&p_sort=&p_desc=DESC&p_direction=Next&p_show=20

Index

Note: Page numbers in *italics* and **bold** refers to figures and tables respectively.

t

u

v

w

Printed and bound by CPI Group (UK) Ltd, Croydon, CR0 4YY

17/09/2026

14964235-0003